Marketing for Architects and Engineers

Marketing for Architects and Engineers:

A new approach

Brian Richardson

Spon Press
an imprint of Taylor & Francis
LONDON AND NEW YORK

First published 1996 by E & FN Spon

Published 2004 by Routledge
2 Park Square, Milton Park, Abingdon, Oxon OX14 4RN
605 Third Avenue, New York, NY 10017

Routledge is an imprint of the Taylor & Francis Group, an informa business

First edition 1996

Typeset in 11/13pt Gill Sans by Acorn Bookwork, Salisbury

A catalogue record for this book is available from the British Library
Library of Congress Catalog Card Number: 96–68679

ISBN 13: 978-0-419-20290-5 (pbk)

Contents

Foreword

Dr Francis Duffy, Chairman DEGW, Immediate Past President of the RIBA

One of the most encouraging indicators of the recent recovery of confidence in British architects in the relevance of their own special skills is their new willingness to use marketing concepts and techniques to help them understand the people, the clients, the society for whom architectural invention is potentially so very important. It was very different not so long ago. Influential architects felt very strongly that to indulge in marketing came very close to selling your body, if not indeed your soul. That marketing means creating the climate of opinion within which architectural skills could be deployed to best effect was, to say the least, not widely understood.

Brian Richardson's book is very timely. It brings together a great deal of hard won experience derived from the better British practices that have survived the hell-raking rigours of the early 1990s. Such practices now know very well what marketing means. The book also benefits from the RIBA's Strategic Study which, besides providing a mass of data, unequivocally insists upon the responsibility of architects to explore with clients, sector by sector, how their requirements are likely to unfold into the next century. If that isn't marketing, I don't know what is.

However, good advice on marketing to architects is necessary but not sufficient. Theory helps a lot too. What is most remarkable about what Brian Richardson has done is that he has been able to turn to theoretical advantage his practical experience in marketing architectural practices. Brian Richardson's ideas on 'synthesis marketing' represent, to me at least, an important advance in marketing thinking in the widest sense because of his insistence on ongoing, creative, value-laden relations between service providers and clients, what he calls 'synthesis marketing'. Is it going too far to claim that it could only have been in the culture of architecture that such a responsible, sympathetic and fresh view of marketing could have been engendered?

Architecture is a profession: architects have voluntarily contracted to work together through their professional institute, the RIBA, to share experience and ideas to help make them individually and collectively better architects. We are trying very hard to modernize the RIBA, to make it a learning, as well as a learned, Institute, using information technology to make the benefits available to all. Advanced thinking on marketing is very much part of this process. What is most gratifying to me is that Brian Richardson's book will benefit not only architects in their individual practices, looking after their own and their client's interests, but also architects acting collectively through the RIBA, foreseeing what benefits architecture should be bringing, through time, to society as a whole.

Acknowledgements

As I give my thanks to those people who have provided assistance and support in the production of this book, I reflect not on a completed manuscript but rather on a period several years before when there was only an inkling of a new marketing discipline for architects and engineers. At this time, Helen Elias, then RIBA regional director, suggested to Caroline Mallinder, commissioning editor, that a book about marketing was much needed and furthermore that I was a possible author. Robin Nicholson, then RIBA Vice President Public Affairs, echoed her sentiments and argued my case. To Helen, Caroline and Robin, I offer my thanks.

I am indebted to Alastair Blyth, *Architects Journal*, for giving me the opportunity to write a series of articles about marketing throughout 1993. I also thank Andrew Rabeneck and Ted Watts, fellow speakers at the 1993 AJ Annual Conference, for their words of encouragement and in particular to Andrew for pointing me in the direction of Robert Gutman and his writing.

In the same year, I wrote to Weld Coxe in the United States. Weld wrote the first book about marketing for architects and engineers thirty years ago and our subsequent correspondence provided a wealth of insight and observation. I would like to thank him for the letters, papers and books that he has sent over the last few years.

On the engineering front, I would like to thank Doug Morton (Hyder Consulting), Mike Brown (Ove Arup and Partners) and Ian Burdon (Merz and McLellan Ltd) for giving up their valuable time and sharing their thoughts. Special thanks go to Doug for reading my script and making detailed comments.

The task of proof reading the book was undertaken by two of my colleagues at FaulknerBrowns, namely Tricia Green and John Ramsay. Their general comments and attention to detail were invaluable. I would also like to take this opportunity to thank all the staff and partners at FaulknerBrowns for being such a great bunch to work with.

William Roe (William Roe Associates) and Richard Clarke (Web 21 Internet solutions) have given me their specialist advice in the areas of scenario planning and the Internet respectively. I would like to thank them for the time they spent explaining these fascinating fields of importance to design professionals in the future.

My thanks go to my fellow board members of Northern Architecture Centre Ltd and our chairman, Geoffrey Purves, for allowing me to use the proposed Northern Architecture Centre as a marketing case study. They had to endure the unnerving prospect that their actions were being recorded for future publication.

And last but by no means least, my thanks go to my wife Carolyn who had to cope with lost evenings and weekends and interrupted holidays during the eighteen-month period of writing the book in my spare time. I would particularly like to thank her for suggesting that Aristotle's concept of 'entelechy' would be a suitable component of synthesis marketing.

To Ruby and Dennis

Introduction

If architects and engineers are to face the challenges and changes of the construction marketplace in the next century, they will need to develop a marketing agenda which can be supported and reinforced at every level. The aim of this book therefore is to develop a coherent and comprehensive marketing discipline that is relevant and applicable to both small and large practices. The intention is not only to improve the day to day marketing function within practices but also to develop a long-term view of organizations and their relationship with the market.

The idea of adapting and adopting existing marketing disciplines used in other fields to cover the provision of design services presents many challenges. Indeed, Neil Morgan [1] feels that:

> the professional service context is a peculiar one. It deserves a book about the peculiarities, irrationalities and idiosyncrasies of professional services marketing as a unique and different type of marketing rather than simply writing about the areas in which marketing is similar in the professional service context to other types of marketing.

In the case of design professionals, i.e. architects and engineers, my view is that their visionary and creative mind set does not respond well to the retrospective analytical tools of traditional marketing. This is especially true when they are asked to reperceive their position in the marketplace. The first challenge is to convince architects and engineers that the adoption of a marketing approach will prove beneficial. In a world of increasing competition, this can only be achieved by demonstrating an improvement in comparative business advantage (the improved performance of a business compared with others) and by showing that they can have more influence over changing events. Each chapter therefore describes why the proposed measures should be taken and what improvements are likely to result.

The techniques of marketing for architects and engineers described in this book are chosen to resonate with the outlook and views of design professionals. For this reason, the two approaches of scenario planning and synthesis marketing are combined to suit the needs and operation of design practices. The aim is to create a vision of what practices will look like in the long term using scenario planning methods and to plan how subsequent relationships with individual customers and the related sectoral infrastructure will develop using a synthesis marketing approach. The sectoral infrastructure is defined as 'every organisation and individual that can influence client perception' (McKenna [2]). The infrastructure of any sector can include trade organizations, trade press, client associations, sectoral institutions, governing bodies or consumer groups. A diagrammatic overview of the marketing process is provided in Fig. 1.1.

In the early 1990s, architects and engineers were having to look carefully at the process of identification and contact with new clients and had to re-examine their relationship with existing clients. As a result the marketing philosophy and approach came into confrontation with the close-held tenets of the other professions. Questions were being asked

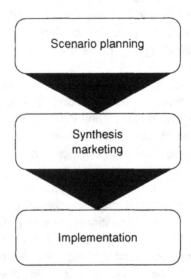

Fig. 1.1 The marketing overview.

on both sides of the apparent divide [3]. How well did design professionals understand the operation of the construction marketplace? What was wrong with the existing, valued relationship that they had with clients? Would the apparent commercialism of marketing and marketers undermine the integrity and standards of professional service provision? What value did marketing have for a practice? Chapter 2 looks at these fundamental issues and identifies the basics of markets and marketing that will be discussed and referred to in more detail in later chapters.

If marketing is to be adapted and adopted by practices, the most significant hurdle to be overcome is the internally generated barrier of resistance to cultural change. This resistance is represented partly by common misperceptions and misunderstandings about marketing but also by a reluctance to adapt to social and economic changes that are taking place globally and within economic regions such as Europe. These misperceptions and misunderstandings are listed and discussed in Chapter 2.

The role and position of design professionals varies considerably in different countries. For example, at the time of writing, almost a half of British architects are either unemployed or underemployed following the longest period of recession since the Second World War. This is in stark contrast to the position in Germany where unemployment among architects is low and architects are afforded a level of protection and status that was accorded to British architects twenty years earlier. However, despite the social, economic and organizational complexities, it is possible to take a long-term view of how and where your practice will market its services. The intention is to describe what you would like your practice to look like in ten years time. Paradoxically, a long-term view is required as an antidote to short-term fluctuation and variation. The long-term view can be achieved by using the flexible and adaptable techniques of scenario planning described in Chapter 3. These techniques are applicable to large and small practices operating at regional, national and international levels.

Building up a long-term relationship with clients, institutions and organizations in a selected market or markets, however defined, is seen as an increasingly important marketing objective. If the aim is to have a stable future for your company, the client base needs to be cultivated, maintained and protected. This is especially true of design service provision in which architects and engineers are engaged in a working relationship with an individual client over a long period of time. A synthesis marketing approach is described in detail in Chapter 4.

In Chapter 5, the strategic options available to practices are outlined and discussed and these options are then mapped out over the time period of the scenario plan. The strategic emphasis here is on choice and empowerment. For example, choosing to win work in specific markets rather than offering a general service to all comers is a strategic choice. The balance between sectoral preference and type of service is just one of a number of important choices. These issues and choices are at the heart of the marketing problem, and there is no quick 'marketing fix'. Every issue needs careful consideration and needs to be addressed in terms that reflect the long-term view of your company and its corporate aspirations.

There would be no point in formulating strategic options for marketing design practices if the strategies could not be turned into results. Chapter 6 takes a looks at the practical aspects of marketing and is seen from the point of view of the person or persons responsible for making things happen.

Chapter 7 looks at architecture centres as a marketing platform for design professionals. This is an example of how to apply scenario planning and synthesis marketing methods in a complex marketing situation. The example draws together all the social, economic and cultural aspects of marketing discussed in earlier chapters. It looks in particular at the use of synthesis marketing methods to formulate a three-point marketing platform of 'culture, community, construction', i.e. the promotion of architecture as a physical manifestation of cultural values, as engagement with the general public, educators, students and the business community, and as support for the construction industry.

Markets and marketing 2

In the 1990s, marketing will do more than sell. It will define the way a company does business. (Regis McKenna [2])

2.1 The marketing function

Christian Grönroos [4] has defined the marketing function as 'all resources that have a direct or even indirect impact on the establishment, maintenance, and strengthening of customer relationships, irrespective of where in the organization they are'. Everyone in the organization has contact with either clients, members of the industry or the general public and therefore all members of staff are part of the marketing effort. It is not desirable that only senior members of the company or a separate unit or person be responsible for marketing if the practice is to be market oriented. Many practices have a system of training young consultants in technical and design matters. They reward selected individuals with promotion and after many years, when they have reached a senior level, introduce them to the marketing issues that face the company. The marketing experience comes too late in the cycle. Young consultants should understand the marketing position of the company and its future marketing plans and aspirations at a very early stage if they are to make a marketing contribution now and become fully effective later in their career.

'A number of key activities that are central to any professional services marketing function' are listed by Neil Morgan [1].

These are:

- researching and analysing the existing marketplaces for the firm's service offerings;
- identifying coherent segments in the marketplace that exhibit relatively similar need;
- analysing the firm's resources, personnel and areas of expertise;
- designing service offerings which translate internal strengths into specific services that meet the needs of particular segments;
- offering only those services and targeting only those market segments that enable the firm to achieve its long-term objectives;
- communicating the service offerings to existing and potential clients; measuring client satisfaction with services and using this information within the firm.

The list is by no means comprehensive but gives some indication of the range of activities. Chapter 4 looks at how these activities should be undertaken within the framework of a synthesis marketing approach and looks at how they vary in the process from initiating contact through to winning repeat business from an existing client. Chapter 6 looks at some of the tactics and techniques that will ensure effective implementation.

In 1992 E. J. D. Warne CB [5] undertook a review of the Architects (Registration) Acts 1931–1969. He undertook the review as an independent assessor. The final comment in the executive summary of his report was that 'The future well-being of the profession should be assured by the talent which it attracts into membership, the long and rigorous education and training process before qualification, the central position which architects occupy in the building process, and above all the increasing awareness among architects of the disciplines of the market, i.e. the need not just to provide quality services but also to provide those services which the client or employer actually wants.' This parting comment embodies the choices and questions that face not only the architectural profession but also all design professionals in the construction industry. What are the disciplines of the market relative to the provision of design services? What are the services which the client or employer actually wants? Is there a marketing discipline for architects and engineers that will answer these and other questions?

In October 1990, the RIBA published market research findings into how practices were addressing the need for marketing (the total survey size was 300 practices; a large practice was defined as having eleven or more full-time architectural staff):

- 43% of practices had someone with formal responsibility for marketing;
- out of the 43%, 36% allocated marketing responsibility to an individual on a part-time basis. The other 7% of practices had a full-time marketing person;
- 49% had advertised their practice in the last year;
- 25% of large practices had a written strategy;
- 11% of all practices had a marketing strategy.

The national recession started after the second quarter of 1990 and was marked by a rapid drop in investment in the UK economy. The RIBA snapshot is therefore taken shortly after the onset of the recession and after the preceding rapid growth of the construction sector in the late 1980s. The research revealed a low level of emphasis on marketing and marketing effort. More than half the practices interviewed were not allocating marketing responsibility to a specific partner or director. Three-quarters of the practices had no written marketing strategy and very few had a formal marketing department. No comparable survey is available for the engineering consultants but anecdotal evidence suggests a similar situation.

So what subsequent steps have design practices taken to increase market awareness and in particular to develop the marketing function? How should the marketing function be initiated and developed in a practice? What scale of marketing is called for? What level of formalization of marketing is required? What should the relationship be between the marketing strategy and the aims of the organization? Or indeed, should practices have any marketing orientation, strategy or effort at all?

2.2 Past and future marketing

It is worth reflecting on what were the characteristics and shortfalls of the past marketing of architectural and engineering services and trying to describe the shape and form of future marketing. Some characteristics of past marketing are listed below:

- poor delivery of services – projects not on time and within budget;
- reactive selling – undertaking work by invitation rather than in anticipation of client requirements;
- lack of adaptability – not anticipating and adapting to the changes that have taken place in key sectors such as health and education;
- poor market understanding – a result of failure to integrate

marketing and market intelligence gathering into the mainstream
activities of the practice;
- lack of focus – trying to be all things to all people;
- marketing technical capability to clients with little technical understanding or interest;
- poor market positioning – an inability to establish a distinct position
in the marketplace.

These are just some of the many results of a lack of attention to the
marketing discipline by both architects and engineers: the corollary is the
failure of the marketing profession to provide a relevant and meaningful
marketing discipline. The characteristics of future marketing might include:

- client consciousness – a better understanding of clients' needs and
perspectives;
- market orientation – understanding and anticipating changes in the
market;
- customer service – making everyone in the practice understand that
they have a part to play in the ongoing contact with the client and
their experience of your service;
- deeper experience in fewer markets – carving out a niche in
selected markets;
- producing a marketing plan, preferably using the scenario planning
and synthesis marketing methods outlined in this book;
- wider service provision – addressing the market with a range of
services and not just a core service.

Again, the list identifies just some of the possible characteristics of future
marketing. A clearer picture of future marketing will emerge from the
following chapters.

2.3 The marketing discipline and a 'crisis of abundance'

At the beginning of the century, mass production methods operated in
the manufacturing sector. Initially, the supply of goods was met by an
apparently inexhaustible demand. Indeed, in 1913, Gide [6] wrote in his
book *Cours d'économie politique*: 'judging by the poverty of the great
majority of mankind, it would seem as if production always lagged behind
wants, and as if all our efforts should be turned towards hastening it on as
much as possible'. Gide felt that a 'crisis of abundance' could be cured by
abundance. It was in the interests of everyone that production should be

'as abundant and varied as possible'. The related law of markets formulated earlier by J. B. Say stated that 'the greater the number and variety of
products, the more markets will there be for each one'. The belief in the
law of markets lasted through both world wars until the early 1960s. In
the 1960s there was a 'crisis of abundance' in consumer goods. Supply
had at last met demand. Manufacturing companies tried to sell products
using the latest selling techniques. The focus of the effort was to sell
existing products using increasingly aggressive selling and promotional
effort. In the late 1960s there was a view that in order to sustain existing
markets and find new markets, the focus had to be on the actual and
potential needs and wants of the customer. The sophisticated and
complex discipline that developed as a result was called marketing.

The discipline of marketing developed therefore in fast moving
consumer goods markets as a result of the 'crisis of abundance' and was
adapted subsequently to suit companies that had individual clients rather
than a large consumer base. The latter discipline was called industrial
marketing and later became known as business to business marketing. In
the late 1980s, professional services in the British construction industry
also had a crisis of abundance. Fig. 2.1 shows the considerable growth

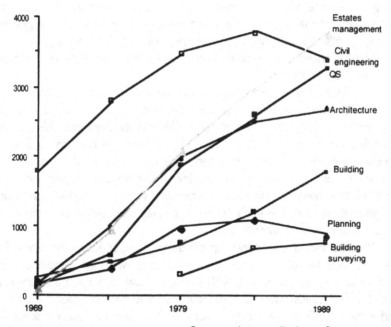

Fig. 2.1 Entrants to degree courses. Source: A compilation of government,
RIBA and CNNA statistics.

in entrants to degree courses in a range of professions in the construction industry. All professions have shown a growth in numbers since the early 1970s.

There is simply an overprovision of qualified people in professional services in the construction industry. This was highlighted by the economic recession that started in 1990. The recession was the largest since the Second World War and brought home the fact that, despite institutional protection and regulation, professional services would have to reassess their position in the market place and would have to develop a relevant marketing discipline if they were to deal with their own 'crisis of abundance'.

There are two schools of thought about the oversupply of professionals. One view is that the people trained as architects or engineers have transferable skills and will be able to find work in other areas of the construction industry or other service sectors. The resulting transfer and promulgation of views and values is seen as a benefit to the respective professions. The other view is that the availability of a large pool of labour undermines the position of those people in full- or part-time employment as architects or engineers. In my opinion, the former view takes no account of the economic and social changes that have been taking place since the 1970s, in particular the casualization of labour, the deregulation of markets and the structural and organizational changes that are taking place within and between professions. (The social, economic and cultural changes in a post-Fordist economy are discussed in more detail in Chapter 3.)

The idea of a marketing discipline for architects and engineers has been met with suspicion, ambivalence and in some cases outright rejection by a limited number of people in both professions. There is also a parallel, wider concern about what Will Hutton [7] calls 'marketisation', namely 'government efforts to extend the operation of the market in a plethora of areas ranging from housing and pensions to television and sport'. Hutton feels that 'instead of extending choice, individual liberty and the general welfare, there has been a secular increase in risk, insecurity and cultural impoverishment'.

Markets based on financial profitability have serious limitations. For example, they fail to recognize the needs and wants of people on low incomes and they cannot deal with social, cultural, ecological and environmental issues unless they are reflected in the balance sheet. However marketing disciplines are being developed that are not guided solely by economic imperatives. For example, in his book *Marketing in Local Government* [8], Kieron Walsh notes that *manufacturers had to respond to the demand of the market because they could no longer sell whatever they made but had to try to make what they could sell. The development of a more sovereign consumer meant that manufacturing industry had to devote increasing energy to finding out what it was that people wanted. As local authorities come to provide more sophisticated and varied services in a more complex environment, so they, too, will need to develop the skills of marketing.*

Walsh also calls for the 'search for a public service orientation which will embody a revised and renewed expressions of the values that underpin local government'. He has therefore attempted to develop a marketing discipline for local government that is not a result of either marketization, commercialization, or an attempt at privatization. It is argued that a marketing discipline for architects and engineers can be developed which is similarly not determined by these factors and does not lead to cultural impoverishment. Architectural and engineering professions will also need to develop revised and renewed expressions of the values that underpin their institutions and practice.

2.4 Marketing and professionalism

In 1988, Robert Gutman [9] wrote that 'What architects resent most about marketing programs is the assumed implication that architecture is a business enterprise rather than a profession, and that the business side is taking precedence and guiding the definition of the field.' The debate about professionalism, commercialism and the role of marketing is ongoing and any meaningful marketing discipline for design professionals must recognize the complex and sometimes contradictory stances, concerns and aspirations that lie behind the arguments. For example, the architectural profession appears to have an ambivalent stance to membership of the construction industry [10]. The profession aspires to represent both the interests of the client, the end user and society as well as taking a lead role in the construction process. Indeed, lack of self-interest, and concerns for the wider community and building quality, are often seen as measures of professionalism. The debate therefore is about whether or not this type of votive professionalism is outmoded at a time when the combined impact of changing procurement patterns, privatization, the use of information technology, and role redefinition within multidisciplinary teams is forcing architects to be specialist service providers on the supply side of the industry. Indeed, Stephen Greenberg [11] has posed the question 'whether architecture hasn't for the foreseeable future switched from a professional to a commercial enterprise'.

It is suggested that the professional ethos of all design professionals should be supported by a complementary marketing agenda at both institutional and corporate levels and that the two are totally compatible. The reason for making such a suggestion is that marketing manages and improves communication with the client. Effective communication with clients can only be in the interests of design professionals and their respective professions and is a valid and necessary component of professionalism.

2.5 Core, actual and augmented services

Many of the strategic and tactical marketing problems that face design professionals revolve around the question of what level of service is required to win the commission and ultimately to satisfy the client. Design professionals offer a design solution to a client's problem. This is the unique core service on offer to clients (see Fig. 2.2). The delivery of the actual service might involve a wider range of other activities than simply design. For example, other disciplines could be involved as part of team, or other duties will be taken on such as contract supervision and management. These additional activities and functions that help to perform the task make up the actual service. In addition, practices might offer, for example, professional indemnity insurance, total quality management programmes, master planning services, collateral warranties, pre-commission feasibility work (possibly on a speculative basis) or advice to the client on maintenance issues after completion of the commission. All these additional services make up the augmented service. The level of

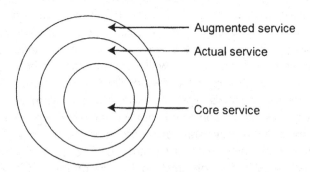

Fig. 2.2 Service analysis. Source: Adapted from Bayliss [12].

service and the difference between quality work and quality service are examined later in the chapter.

2.6 A marketing philosophy

Regis McKenna [2] was quoted at the beginning of this chapter. He said that 'In the 1990s, marketing will do more than sell. It will define the way a company does business.' The way of doing business is the embodiment of the marketing philosophy and is marked by the following nine-point approach:

1.   Listen to clients.
2.   Understand the client's perspective.
3.   Manage the marketing effort.
4.   Develop a long-term vision and be consistent for success.
5.   Match quality work with quality service.
6.   Cooperate as well as collaborate.
7.   Influence as well as control.
8.   Adopt a simple, systematic and sustained approach to marketing.
9.   Market to the sectoral infrastructure not just the client.

2.6.1 Listen to clients and understand the client's perspective

The first task in marketing is simply to listen. Active listening is a skill that can be a natural gift but is more often taught. It is the art of reflecting positively on what is being heard without overriding the client's view because you feel, for example that his technical knowledge is limited or that her views are out of date or any of the many other reasons that come to mind when you are listening to the client. As well as noting the client's view of the proposed project, look for the hidden agenda. For example, is this the first commission for which this individual client has been responsible? Does this commission have any bearing on the client's career prospects? Is funding a problem? By listening actively you are beginning to understand the client's perspective. That perspective could reach well beyond the practical details of the commission and you must make sure that the client is aware of your understanding of his or her situation.

2.6.2 Manage the marketing effort

The traditional view of marketing and the marketing function, as illustrated by Christian Grönroos [4], is shown in Fig. 2.3. The upward arrow

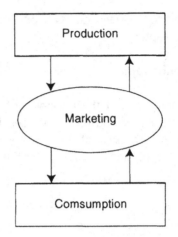

Fig. 2.3 The traditional role of marketing. Source: Grönroos [4].

'indicates that the marketing specialists acquire information about the market'. The downward arrow, on the other hand illustrates the 'planning and executing of marketing activities'.

The alternative synthesis marketing approach, recommended for design practices, is shown in Fig. 2.4. Arrows to the right indicate strategic and tactical marketing input and effort. Arrows to the left show marketing reviews and adjustments. The initial marketing strategy that preceded contact with the client is not a tablet of stone and changes are made throughout the contact with the client. Over a period of several years, there could be substantial changes to the client's situation, the services that your company offers or the economic backdrop. Chapters 4 and 5 will look closely at the management of the marketing effort within a synthesis marketing framework.

2.6.3 Develop a long-term vision and be consistent for success

In Chapter 3 I will outline how you can create a vision of what your practice will look like in ten years' time and its operation in the ensuing period. The marketing argument for adopting such an approach is that vision and consistency lead to success even though events may change dramatically in the ten-year period. Research and observation has shown that the professional leaders of successful practices have a consistent philosophy and approach to the way a firm does its work. Several important observations were noted by Coxe et al. [13]:

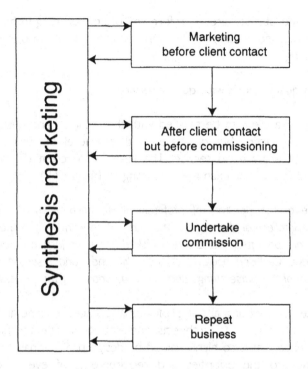

Fig. 2.4 Synthesis marketing

- 'Individual firms that have the longest records of successful achievement appear to do what they do in a fairly uniform manner. In other words their approaches to doing work and operating the firm are clear and these practices are followed consistently over time.'
- 'The area in which the consistency is most notably different from firm to firm – but in which consistency within a given firm seems to yield important results – is in project delivery systems, that is, how the work of the firm is done.'
- 'In organization terms, the firms that do best tend to adopt one approach to their leadership-ownership structure and do not tamper with it from generation to generation.'
- 'For many firms the course of greater wisdom is to reach for some consistency between its practice areas so that a single approach to management practice can be devised. Such consistency need not be absolute, but it must be sufficiently clear to provide a reasonably stable centre of gravity.'

Consistency and stability are therefore important for long-term success and will be discussed in Chapter 3 in the context of scenario planning and superpositioning.

2.6.4 Match quality work with quality service

It is possible for a client to be satisfied with the technical and design input of a particular practice but dissatisfied with the level of service. Good work does not imply good service. The reasons for client dissatisfaction are many and varied. For example, according to Maister [14]:

> Because of the proclivity of professionals to become more fascinated with the intellectual challenge of their craft than with being responsive to clients, all too often clients are mocked for their lack of professional knowledge, despised because of their demands, and resented because they control the purse strings and hence autonomy of the professional.

David Maister was writing about professional services in general but the comments apply equally to architects and engineers. The technical and design excellence that is so highly valued by design professionals may have little relevance to the priorities and requirements of even the most experienced client. Good client satisfaction comes from managing the client's experience of your service as well as undertaking the work.

The rapidly changing patterns of procurement that are a feature of the 1990s have led to the proliferation of clients with limited experience of building and design services. If architects and engineers are to become more successful in the marketplace, they will have to learn to deal with these inexperienced clients and their service requirements. Inexperienced clients exhibit the following characteristics:

- have had limited contact with the construction industry;
- do not understand the nature and merits of the design service that is on offer – indeed, the qualitative and quantitative contribution of good design is hard for even the most experienced client to appraise;
- have no method of comparing one practice with another and, if in doubt, have a tendency to assume that they are comparing like with like;
- find difficulty in drawing up the initial brief for the project in construction and design terms;
- have no idea of the timescale for development and the problems

that are likely to occur, have unrealistic expectations of the service that consultants provide.

So how can architects and engineers deal with the inexperienced client and ensure satisfaction with the service provided? Part of the solution lies in agreeing the appropriate level and type of service that is required. For example, a client might be concerned about the requirement for low running and maintenance costs for a particular building and might articulate that requirement totally in terms of the augmented service. The marketing task is to show that good design will produce these outcomes and will satisfy the client's requirements. In other words, the design professional has to redefine the work in terms of the core service. The process of negotiation and discussion is part of the actual service that is on offer to the client. Dealing with the client's perceptions and expectations, however unreal they may be, is also part of the actual service. The skill lies in adjusting the level and type of service, i.e. core, actual and augmented, to meet the client's requirements. The consultants' tendency to relate everything back to the core service will only lead to client dissatisfaction if the core service has not been agreed and understood by the client.

2.6.5 Cooperate as well as collaborate

In May 1992 the RIBA published *Phase One: Strategic Overview*. This report was the first part of the strategic study of the profession [15]. 'The objective was to lay the foundations of a new Institute strategy to help all architects – whether principals, or assistants, employed in government or local government, consultants, teachers, or architects working in any other capacity – to prepare themselves to practice architecture more effectively in a rapidly changing world'. One of the papers was by Andrew Saint, who wrote:

> It is fair, then to say that most architects prefer collaboration to competition; Kropotkin, not Adam Smith, would be their favoured economic philosopher. Construction is, after all, a collaborative business; and although architects have a reputation as the prima donnas of the industry, less willing to compromise or adjust than other building professionals, we must assume that they expect and intend to collaborate. The need for collaboration is hardly ever at issue; more contentious is its nature and quality.

The requirement to collaborate and be part of a team is not in question. What is in question, particularly in the case of architects, is the intention and desire to cooperate. Architects, in particular, are having to face up to the fact that they will not always be the team leader in any collaborative grouping. Design professionals will have to learn to cooperate as well as collaborate.

2.6.6 Influence as well as control

Influence is the capacity to modify the perception of others. Control is the capacity to restructure situations. In the marketing arena, architects and engineers have relatively little control. They have to refine the process of observation, anticipation and influence and go as far towards the centre of Fig. 2.5 as quickly as they can. At the professional level, architectural and engineering institutions are having to anticipate and influence events in larger numbers and with greater frequency. These events can range from the proposed changes to procurement methods for the construction industry to deregulation of service provision. At the corporate level, companies are having to re-examine their relationship with the market-place and to see what influence they can exert. The marketing discipline for architects and engineers is concerned with the maximization of influence on clients, the sectoral infrastructure and the construction industry.

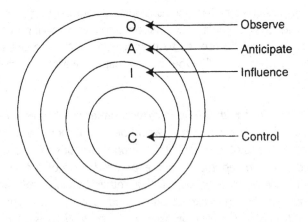

Fig. 2.5 The CIAO diagram

2.6.7 Adopt a simple, systematic and sustained approach

The solution to the problem of marketing in a changing environment is to adopt a simple, systematic and sustained approach to both strategic and tactical marketing (see Fig. 2.6).

Marketing is not a quick fix solution to dealing with the peaks and troughs of practice workload. If your practice workload drops and you suddenly decide to go out and win work to cover the shortfall in the next few months, you have left your marketing effort too late. Marketing has to be a constant, integrated part of your agenda, no matter how big or small your practice. You have to allocate a portion of your time to marketing on a regular basis and you have to spend time reviewing your strategy.

2.6.8 Market to the sectoral infrastructure as well as the client

As stated previously, the sectoral infrastructure is defined as every organization and individual that can influence client perception. The infrastructure of any sector can include trade organizations, trade press, client associations, sectoral institutions, governing bodies or consumer groups. If your practice is to be successful in the long term, it will have to market to the sectoral infrastructure as well as to individual clients. You have to ensure that your practice is accepted by everyone as a serious player in that particular market. Marketing to the sectoral infrastructure is discussed in more detail in Chapter 4.

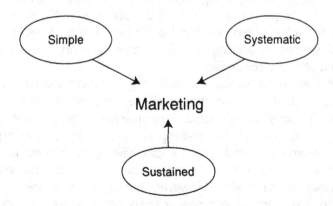

Fig. 2.6 Simple, systematic and sustained marketing

2.7 Misperceptions and misunderstandings

There are a number of common misperceptions and misunderstandings about marketing that need to be noted and addressed before any coherent and meaningful explanation of its true nature can be given. One misperception has been mentioned already, namely that only senior members of staff are involved in marketing. Another common misperception is that marketing and selling are the same. According to Theodore Levitt [16], 'selling is preoccupied with the seller's needs to convert his products into cash, marketing with the ideas of satisfying the needs of the customer by means of the product and the whole cluster of things associated with creating, delivering and finally consuming it'. Selling is concerned with taking existing off-the-shelf goods or services and finding a buyer. Marketing is concerned with responding to client need. Levitt's traditional account of the difference between selling and marketing uses the term 'product'. The word service can be interchanged with the word product but there must be some understanding of what is meant by service. The difference between core, actual and augmented services was discussed earlier in the chapter.

Another common misconception is that marketing is about advertising and public relations. In fact, these activities are only one small part of three major areas of marketing, namely 'marketing services', 'service marketing' and 'internal marketing', described earlier in the chapter. This misunderstanding stems in part from the relatively high visibility of the advertising and public relations functions and also from the limited application of the marketing discipline in the past.

Many people believe that marketing ends when a commission is won. The application of strategic and tactical marketing effort should not be limited to the period up to winning the commission. Marketing should play a part in a longer-term relationship. Chapter 4 offers a detailed explanation of synthesis marketing in which marketing effort is made before, during and after the commission.

It is also untrue to say that only large practices should undertake marketing, or that the marketing discipline does not apply to smaller practices. The only differences are the manner in which marketing is carried out and the marketing challenges that face each type of organization. For example, a small practice can have good internal marketing communication and can therefore respond well to changes and opportunities in the market. However, it may not have the resources needed to make inroads into new markets or foresee market developments.

2.8 The level of formalization

All marketing activities require a level of formalization within the organization. The term 'level of formalization' is used to mean the procedures, meetings, written records and documentation covering the marketing effort and the degree to which it is organized and structured. It is possible, for example, that there could be a considerable level of marketing effort in terms of human and financial resources but a low level of formalization. Similarly, formalized marketing does not ensure effective marketing effort.

There is a limit to the extent to which design practices can adopt a more formal marketing approach. The limit is due to both constraints on resources and the understanding that design professionals have of the marketing discipline. Indeed, formality might lead to rigidity and could be completely undesirable if the practice is to remain responsive to market needs. It is suggested therefore that formalization is kept to a minimum but with a clear understanding that a system of marketing reviews involving written records be agreed and adhered to. The review system described in later chapters is intended to combine brevity with clarity.

2.9 Internal marketing, marketing services and service marketing

A distinction should be drawn between service marketing, marketing services and internal marketing. Examples of the marketing activities in each area are shown in Fig. 2.7.

2.9.1 Service marketing

Service marketing is concerned with all aspects of direct contact with the potential client. It requires a good understanding of the marketplace and a clear identification of the client's decision making unit, i.e. the group of people invested with the necessary purchasing power. The process of service marketing involves understanding the needs, initiating contact and developing an ongoing relationship with the decision making unit.

2.9.2 Marketing services

The term marketing services covers all marketing effort targeted at the sectoral infrastructure. As stated previously, the sectoral infrastructure [2]

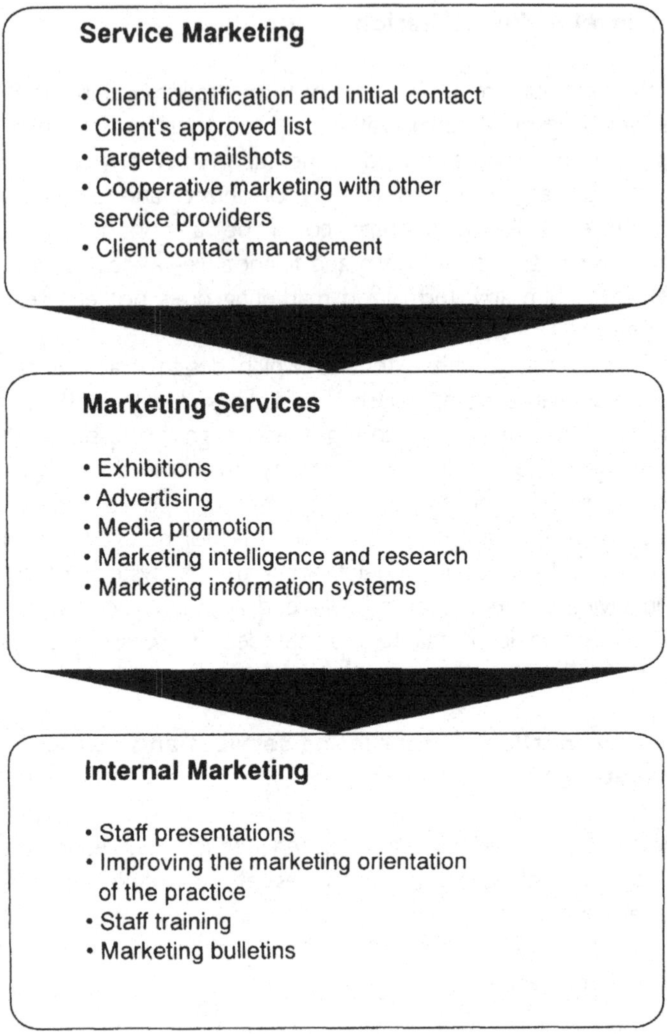

Service Marketing

- Client identification and initial contact
- Client's approved list
- Targeted mailshots
- Cooperative marketing with other
  service providers
- Client contact management

Marketing Services

- Exhibitions
- Advertising
- Media promotion
- Marketing intelligence and research
- Marketing information systems

Internal Marketing

- Staff presentations
- Improving the marketing orientation
  of the practice
- Staff training
- Marketing bulletins

Fig. 2.7 Service marketing, marketing services and internal marketing

is defined as every organization and individual that can influence client perception. The infrastructure of any sector may include trade organizations, trade press, client associations, sectoral institutions, governing bodies or consumer groups. For small practices operating within a regional economy, the infrastructure may also include key organizations and institutions that influence the regional economic network.

2.9.3 Internal marketing

Internal marketing covers the communication of marketing effort and understanding within the practice. Marketing communication and marketing effort begin internally. Good internal marketing makes staff aware of their marketing role and leads to effective implementation.

The reason for drawing the distinction between service marketing and marketing services is that many practices see their marketing effort solely as improving promotional activity. For example, practices may see their marketing agenda as a choice between increasing advertising spend, sending out press releases and generally adding weight to their marketing support services. After a few months, questions are being asked about the benefit and effectiveness of an increase in marketing service effort. Partners or directors begin to realize that they have to face the far more complex and difficult task of service marketing. In other words, they have to readdress the marketplace.

2.10 Market reach

Market reach is a result of intellectual transfer capability, potential customer base and available level of technology. In the case of architects and engineers, the market reach is global. The level of available technology is high, the transfer of design ideas is easy with modern communications and there is an emerging global customer base. The only real barrier to market reach is cultural resistance and misunderstanding. Technical and linguistic barriers are relatively easy to overcome.

There is no implication or assumption in the statement about market reach that global operation should be the norm or even desirable. Global market reach provides potential opportunity rather than automatic strategic choice. The point is that if you operate at national or regional level you are choosing not to exploit market reach. Your competitors from outside your region or from other nations may not take the same view. You might face competition from new directions and in novel ways.

2.11 Total quality management (TQM) and marketing

Total quality management is defined by Bovée and Thill [17] as 'an organisational philosophy based on the pursuit of quality and the management practices that lead to total quality'. TQM is a holistic approach that requires quality service provision to clients with a corresponding quality culture throughout the entire practice. Quality is defined as conformance to requirements and is seen as a necessary ingredient for long-term success.

TQM should not be confused with accreditation to BS EN ISO 9000 or any other quality standard. Total quality management programmes extend beyond the basic requirements for accreditation. At the time of writing, many architectural and engineering practices have quality accreditation but few have a TQM programme.

According to John S. Oakland [18], head of the European Centre for Total Quality management at Bradford Management Centre, 'quality starts with marketing'. 'Marketing establishes the true requirements for the product or service. These must be communicated properly throughout the organization in the form of specifications.' Marketing is seen therefore as the process through which requirements are defined and communicated to the organization and is at the leading edge of TQM implementation.

It is suggested that in the case of architects and engineers, the combination of client expectations, perceptions and experience of the service should be substituted for the term 'client requirements and specifications'. The unique contribution that architects and engineers make in the design process goes beyond a simple definition of specifications and requirements. Figure 2.8 provides a diagrammatic representation. Client expectations, perceptions and service experience are three key strands of synthesis marketing and are discussed in more detail in Chapter 4.

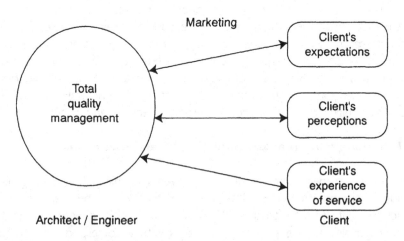

Fig. 2.8 Total quality management and marketing

2.12 Marketing and culture

Organizational culture is the set of collective values, standards and assumptions that determine how a practice will respond to events and situations. It operates at two levels. The first level is reflected by the signs, symbols and statements that send a message about the practice to the outside world. This message is called the corporate identity. The second level of organizational culture is the combination of subcultures and countercultures that exist in even small practices. Subcultures and countercultures are a necessary part of a dynamic, productive practice and have to be addressed by the internal marketing programme. In the case of architects and engineers, there is also a wider cultural agenda to be considered, namely the promotion of ideas and views about the quality and form of the built environment. Figure 2.9 provides a representation of the three cultural levels.

Traditionally, marketing has only been associated with corporate identity, i.e. with only one of the three cultural levels. However, as stated at the beginning of the book, there must be a marketing agenda that can be supported and reinforced at every level. Within each practice, internal marketing must address the interests and concerns of sub- and counter-cultures as well as playing a key part in the total quality management programme and the facilitation of change. Outside the practice, service marketing and marketing services must provide comparative business advantage as well as sending out messages about the practice and its values.

In the wider cultural arena, a marketing discipline must be developed to promote social and cultural values as well as delivering economic returns. Chapter 3 discusses how marketing has developed in response to

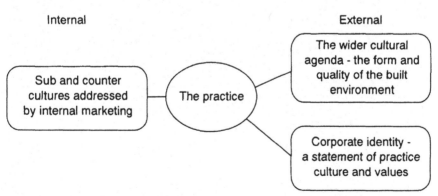

Fig. 2.9 Marketing and culture

post-Fordist economic imperatives. The result has been a proliferation of the incorrect view that marketing is synonymous with commercialism. In the past, marketing has been used mainly for commercial ends and has been defined in terms of profitability. However, marketing disciplines are being developed to address social and cultural markets in which profitability is no yardstick.

Scenario planning

Scenario plans are "a set of organized ways for us to dream effectively about our own future'. (Peter Schwartz [19])

3.1 Scenario planning explained

In order to ensure success, a practice must develop a view, a vision, of what it will look like in ten years time. Scenario planning is the suggested approach to developing such a vision and is, according to Peter Schwartz [19] 'the art of the long term view':

> *Scenarios are stories about the way the world might turn out tomorrow, stories that can help us recognize and adapt to changing aspects of our present environment. They form a method of articulating the different pathways that might exist for you tomorrow, and finding your appropriate movements down each of these possible paths.*

At first sight, the idea of taking a ten-year view seems preposterous. How can anyone take such a long-term view of anything? Design practices exist in a complex, ever changing society and in a construction industry vulnerable to economic recession. And within practices there can be conflicts and dynamics that would appear to be unsustainable for even three or four years let alone ten. So why try to take such a long-term view?

A long-term vision combined with consistency are a common element of many successful practices. The advantages of adopting a scenario planning approach are as follows:

27

- The scenario planning approach throws up a wider array of options and possibilities than would have been revealed using more traditional marketing and planning methods.
- Scenario planning methods bring into play imaginative and creative thinking and limit counter-intuitive blocks.
- The process highlights interconnections that may not have been evident using other methods.
- Later decision making is made easier because you have already crossed many bridges in the thinking behind the scenario planning and in the process of formulating a scenario plan for your practice.
- Scenario planning is a way of identifying critical success factors. Critical success factors are the set of factors which if achieved satisfactorily will result in the overall success of the venture.

Scenario planning is about dealing with uncertainty. It is about stating clearly who you are, what you want to be and how you want to get there. Economic fortunes and social upheavals come and go, but scenario planning should be flexible enough to account for variation and clear enough to provide the necessary direction.

Scenario planning is the formulation of a vision of your practice and a statement of your aspirations set against a debate about the social and economic changes that are likely to occur within the ten-year period. The resultant marketing policy is incremental. It will continue to be changed throughout the lifetime of the organization and will be adapted to suit changing desired objectives.

In the early stages of the process, there is no framework within which plans are drawn up. There is however a set of broad statements about the practice with open ended possibilities for direction. Engineers, in particular, may find difficulty in adapting to this way of thinking in the early stages. The initial process is unlike the problem solving, convergent engineering discipline but is more akin to the creative design process. There is a requirement for a 'willing suspension of disbelief' and a need to tell stories rather than find solutions. The power and benefit of the approach is demonstrated in the later stages of the process. Complex, sophisticated agendas can be developed in a short space of time using scenario planning methods. Figure 3.1 provides an outline of the process.

3.1.1 Identify the trigger

The first step to be taken is to identify the trigger that has made you and your colleagues want to re-examine your view of the practice. The

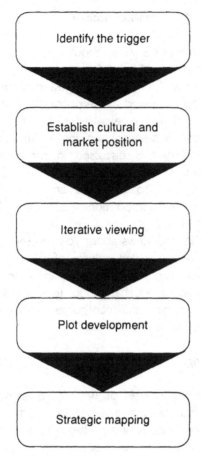

Fig. 3.1 The scenario planning process

reasons could be many and varied. For example, there could be an underlying feeling that the practice is not performing well or that there is a general lack of direction. Similarly, the appointment of a new director or an enlargement or contraction of the practice might lead you to think again about who and what you are. The first step therefore is to do some work on why you want to take a longer-term view. At this stage you have a choice of doing the work in house or bringing in an outside, skilled facilitator who is able to offer an independent voice. None of the procedures and processes described in this chapter demands external facilitation – it is merely a step you might wish to examine if you antici-pate difficulties or if you feel that you need outside help for whatever reason.

3.1.2 Establish cultural and market position

The second step is to express views about where your practice is at the present moment. The starting point for reperceiving your practice is to say where you are now. One of the suggested methods of articulating your present position is called 'superpositioning'. It is an approach developed by Coxe *et al.* and is described in their book *Success Strategies for Design Professionals: Superpositioning for Architecture and Engineering Firms* [13]. Superpositioning is discussed in more detail later in the chapter. It is just one of a number of ways of stating your current position in terms relevant to architects and engineers. There is no compulsion to adopt the superpositioning approach. All that matters is that you find a way that suits your practice.

It is suggested that the views about your present position should be stated in terms of the perceived culture of your organization and its market position rather than technical and management issues. It is very easy to get bogged down in a debate about topical technical and management matters. These technical and management discussions should be put on hold until a later stage in the process.

3.1.3 Iterative viewing

The next stage is to ask a series of 'what if?' questions. There are two aims: first to identify the existing untapped or rejected opportunities that exist for your practice, and second to begin to look to the future. For example, let us assume that your practice has operated within one geographical region in the United Kingdom. You want to reappraise the geographical limits. The first 'what if?' might be the possibility of operating nationally. You decide that that this is not possible or desirable and look at the reasons why. You then look at the option of operating within a reduced region. You decide that the limits are too narrow and again note the reasons. The process is repeated and you narrow the oscillation of the 'what if?' scenarios. What if you operate in an expanded region but with national coverage in selected market sectors? And so on. This is an iterative process of re-evaluation of boundaries. You are reperceiving your practice in the context of wider opportunities and limitations.

The process provides a broader interpretation of your current situation and is not limited to a statement about your current activities and organization. The fuller picture of latent possibilities is related to the constraints and limitations that have been placed on the firm by your own attitudes and views as well as by external considerations. It is only by recognizing the limitations that you have placed on your practice and relating these to the latent possibilities that you will begin the process of empowerment and practice development.

In the case of territorial coverage, the issues might be straightforward. However, the discussions could be more complex when taking a fresh look at, for example, design or cultural issues. Putting existing, long-cherished values and standards into the arena for debate can be both challenging and threatening to many practices. Whoever is responsible for the marketing effort within practices should recognize that he or she will have an increasing role in persuading, influencing and facilitating change in the next decade. Scenario planning is only the starting point in a long-term programme of change management.

3.1.4 Plot development

You are now going to tell a few stories about your practice that describe what it will look like in ten years' time. Keep it simple at the beginning. Start with a few adjectives describing your vision and write the words up on a board for all to see. Ask individuals to talk about what they mean by each word and what lies behind the statement. Peter Schwartz [19] notes that 'scenarios often (but not always) seem to fall into three groups: more of the same, but better; worse (decay and depression); and different but better (fundamental change)'. Limit the number of stories that you tell to between two and four. If you tell more than four stories, the complexities and complications will become unmanageable and the impact of the process will be diluted. Telling only a single story is too constraining and smacks of finding a 'solution' rather than examining possibilities. Scenarios work because 'people recognise the truth in a description of future events'. There is no suggestion that the truth is singular.

At first, story telling might not be easy. The aim of scenarios is not to predict the future. The intention is to 'perceive futures in the present' and reperceive the present in the future. Apart from an initial reluctance to expose your views to your colleagues and superiors in the company, the two main blocks to plot development are denial and dominance. I believe that everyone has an in-built talent and desire to say something about his or her future and to contribute to the scenario planning exercise. What will stop them is a denial that anything is wrong and the dominance of a key individual or subgroup who are only interested in scenarios that reflect their own views.

There are three building blocks [19] for scenario plans 'that give structure to our exploration of the future', and they are shown in Fig. 3.2.

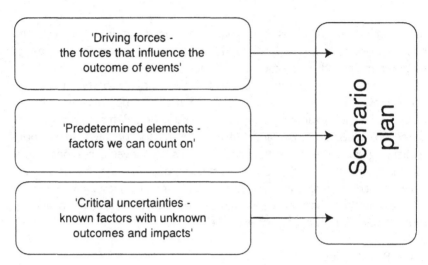

Fig. 3.2 The three elements of scenario planning

The way to identify the driving forces, predetermined elements and critical uncertainties of your world is to reflect on what has happened to your practice in the last ten years. If your firm is younger, you can chose a company that is well known to everyone in your group. What has happened in the last ten years? What were the driving forces that made the company what it is today? What could the company have relied on during the period? What happened during the ten years that could not have been foreseen and which seriously affected the company and its organization?

3.1.5 Strategic mapping

Strategic mapping is a way of examining a series of strategic options, selecting those that will deliver your required outcomes and mapping out their implementation in the scenario planning period. An array of strategic options and a description of their relevance and use are provided in Chapter 5.

3.1.6 How long will scenario planning take?

The first two stages of identifying the trigger and establishing the present cultural and market position can be undertaken relatively quickly, i.e. in less than a day. However, iterative viewing, plot development and strategic mapping will take longer. The effort can be spread over an extended period such as a year. People need time to reflect on proposals and to work on elements of the scenario plan. However, depending on the situation, results can be achieved in a much shorter period. Remember that marketing is not a quick fix. It has to be simple, systematic and sustained in order to have an enduring effect.

3.2 Post-Fordism and scenario planning

As stated previously, the formulation of a vision of your practice and a statement of your aspirations should be set against a debate about the social and economic changes that are likely to occur within the ten-year period. Is there any existing debate about future social and economic changes that will help you to recognize the forces and influences on your practice?

There is no shortage of futurologists and marketing and management gurus. Current figures include Alvin Tofler, Peter Drucker, Michael Porter, Philip Kotler, W. Edwards Deming, Charles Handy, Tom Peters and Henry Mintzberg. Their collective thoughts are considerable and wide ranging but is there a coherent, comprehensive view that will draw selectively from current thinking and 'encapsulate a totality of change' [20]? It is suggested that, in part, the post-Fordism debate provides such a view. In the last thirty years management and marketing methods have been developed in response to 'the shift from centralised mass production to diversified markets with decentralised organisational structures and flexible funding' (Hutton [21]). The shift has been called a transition from Fordism to post-Fordism: a transition from one phase of capitalist development to a new phase. According to Professor Ash Amin [20] 'the passing age, with its heyday in the 1950s and 1960s has been named 'Fordism', a term coined to reflect loosely the pioneering mass production methods and rules of management applied by Henry Ford in his car factories in America during the 1920s and 1930s'. The phrase post-Fordism, according to Eric Hobsbawm [22], 'emerged from Left-wing analyses of industrial society and was popularised by Alan Liepitz, who took the term 'Fordism' from the Italian Marxist thinker Gramsci'.

Professor Stuart Hall [23] provides a succinct view of post-Fordism as

> a shift to the new 'information technologies'; more flexible, decentralised forms of labour process and work organisation; decline of the old manufacturing base and the growth of the 'sunrise industries'; the hiving off or contracting out of functions and services; a greater emphasis on

choice and product differentiation, on marketing, packaging and design, on the 'targeting' of consumers by lifestyle, taste, and culture rather than by categories of social class; a decline in the proportion of the skilled, male manual working class, the rise of the service and white-collar classes and the 'feminization' of the work force; an economy dominated by multinationals, with their new international division of labour and their greater autonomy from nation state control; and the globalisation of the new financial services markets linked by the commu-nications revolution.

Although many of Professor Hall's comments relate to the manufacturing sector, his observations have relevance and application to the service sector and the construction industry. The introduction of information technology and the communications revolution has had a major impact on both the supply and demand sides of the equation. The ability to handle large amounts of information across great distances has led to the deregulation and diversification of the capital markets. Instead of concen-trating resources in single large projects, companies are able to obtain funding packages for niche markets. The increased use of information technology has led to increasing specialization in order to meet the complex needs of sophisticated clients and consumers. A number of benefits have resulted from this process. In the construction industry, the increased awareness of client and consumer perspectives has in some cases resulted in the provision of a better service. However, improve-ments to the delivery of services and construction have not necessarily been matched by improvements in design, purpose and value.

A parallel aspect of post-Fordism that is not covered by Professor Hall's summary but is relevant to architects and engineers is the casualiza-tion of labour and its social consequences. Will Hutton [24] suggests that we live in 'a 30/30/40 society':

There is a bottom 30 per cent of unemployed and economically inactive who are marginalised; another 30 per cent who, while in work, are in forms of employment that are structurally insecure; and there are only 40 per cent who can count themselves as holding tenured jobs which allow them to regard their income prospects with any certainty. But even the secure 40 per cent know they are at risk; their numbers have been shrinking steadily for 20 years. The 30/30/40 society is a proxy for the growth of the new inequality and of the new risks about the predict-ability and certainty of income that have spread across all occupations and social classes.

Hutton argues that 'the so-called 30/30/40 society is not only undesirable, it is unsustainable'.

The changes that have taken place have been justified on the grounds of efficiency but according to Hutton 'the promotion of uncertainty, risk and insecurity has made the operation of the economy as a system less efficient. It has weakened the growth and stability of demand; it has reduced firms' incentives to invest in their workforces and their infrastructure; it has inflated current public expenditure and reduced the tax base'.

By the end of the 1990s, I suggest that the majority of architects and engineers will have slipped into the bottom 60% of marginalized and insecure employment despite the upturn in the economy and the projected growth of the construction industry. This is in part due to the oversupply of professional services referred to in Chapter 2 and the continuing post-Fordist labour market restructuring. The reason for the restructuring is that it provides a method of economic regulation through a flexible labour market. The demand side changes and restructuring include the creation of new forms of markets through, for example, the devolution of powers to bodies such as health trusts and universities. These organizations have now become independent consumers of services in which specialist services are on offer to niche markets.

Are there any characteristics of a practice that is more likely to succeed in such an environment? It is suggested that such a practice would exhibit at least two or more of the following:

- The practice would offer consistent, clearly defined, specialist services to niche markets.
- The practice would be accepted as a serious player by the sectoral infrastructure and the nature and extent of the services on offer would be well known and understood.
- The practice is likely to be a market leader. This can mean that the practice has either the largest volume sales in its chosen sector or is ahead of the market by virtue of technical and/or design innovation. This split between the two forms of market leadership and the pressure to move in one of these directions becomes more important when we look at the polarization of the superpositioning matrix later in the chapter.
- The firm has a clear understanding of the level and type of technology appropriate to its target markets.
- The practice has achieved a balance between the extent of its geographical coverage i.e. the extent to which it needs to exploit

market reach, and the size of its customer base required to ensure regular work or meet growth targets.

■   The practice has achieved a balance between sectoral coverage i.e. the number of sectors in which it operates and is considered to be a serious player, and the level of work flowing from each sector that will ensure regular work or meet growth targets.

■   The practice has adopted employment practices and group working methods to suit the supply of design professionals and the selected market sectors.

There are limitations to the applicability of the post-Fordist perspective. For example, no economy could be said to be uniformly post-Fordist. The post-Fordist view also suggests that there was a distinct break between a Fordist and a post-Fordist period. Again that does not stand up to deeper intellectual analysis. The point about using a post-Fordist perspective for scenario planning purposes is that it allows us to put flesh on the story. The intellectual shortcomings of post-Fordism are almost irrelevant. What is important is that imagination is used to develop a set of options and that the assumptions are fully understood. Whether you agree with this kind of post-Fordist perspective or not, from a marketing standpoint you should be aware of:

■   the issues in the debate;
■   the implications for the strategic and tactical marketing of design services in a post-Fordist economy;
■   the changes that are likely to take place in the local and regional economy;
■   the economic and social changes that will determine the nature of employment of design professionals in the future;
■   the globalization of design service provision;
■   the impact on cultural consumption.

3.3 Cultural consumption

According to Robert Gutman [9], 'cultural consumption' takes a variety of forms. The first is in the activities of developers and funders who consume architecture through their appointment of design professionals. The second is the attempt by both local and central government to improve the quality of public space and buildings. And the third is through the increasingly sophisticated demands of the general public who want good design at home, in their leisure space and at work. Many members of the public are attempting to 'incorporate tokens of architectural culture into their surroundings on a scale they can afford'.

Gutman sees cultural consumption as a sign of a mature society in which 'appreciation of aesthetic values is no longer confined to an upper class but extends to many more groups in society'. The corollary is that cultural consumption should involve public participation and should take into account public opinion. Cultural consumption should include public participation in the decision making for new buildings, refurbishment and the preservation of historic buildings. Any scenario plan and synthesis marketing strategy for a practice should take into account increased user and community participation in the early stages of the process. Design professionals will have to learn new skills of cultural marketing in response to changing patterns of cultural consumption and will have to learn to influence as well as control.

From a marketing perspective, the important point to note is that the distinction between culture and consumption no longer exists. Cultural consumption and cultural consumers are firmly established in a modern economy. The issues of cultural awareness and consciousness raising that were prevalent in the 1950s, 1960s and 1970s are now super-seded by the needs and aspirations of sophisticated consumer groups and corporate clients. Design professionals will have to learn how to deal with these demands through marketing as well as cultural measures.

3.4 Globalization

What is globalization, and does it matter anyway? In theory, true globaliza-tion is the free transfer of capital, people and ideas: not a situation that appears to exist worldwide at the moment. However, the existence of global market reach, made possible by related technologies, allows global operation and global networking. If you have a design practice that has offices with permanent staff in at least two continents and employs native people in senior posts within the organization, then it would be fair to say that your practice operates globally. If your practice provides specialist services and is willing to consider collaboration with other services providers in any location worldwide then your company is involved in global networking.

As stated in Chapter 2, there is no implication or assumption that global operation or global networking should be the norm or even desirable. The existence of global market reach provides potential oppor-tunity rather than automatic strategic choice. What should be recognized is the long-term impact of macroeconomic and cross-cultural change on the construction industry. Global interdependence means that social, cultural and economic changes in one country have a direct impact on other countries. One of the catalysts for this effect has been the establishment of global financial markets. A financial deal in one region of the world is echoed by an economic change in another. At the same time, tastes and standards appear to be converging across countries and there is an increasing homogeneity of products and services.

According to Brian J. Lewis [25], the provision of architectural and engineering services will become global in the 1990s (see Fig. 3.3). Lewis feels that 'if you can get off any plane at any international airport and within 45 minutes be eating a McDonalds hamburger cooked to worldwide uniform specifications, can engineering be far behind?' There is no doubt that there are products and services that can demonstrate cross-cultural attractiveness and benefit. British architectural and engineering service provision may well be one of the globally attractive services of the late 1990s and beyond. Indeed there is a great deal of anecdotal evidence to suggest that British architects and engineers are held in high esteem throughout the world. What is more questionable is the willingness and ability of British architects and engineers to market their services on a worldwide basis.

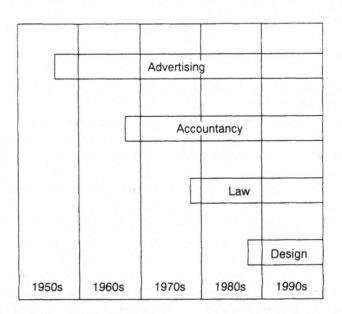

Fig. 3.3 Five decades of globalization of professional services. Reproduced with the permission of the Brian J. Lewis Company

In any international scenario plan, globalization and the development of technology are two of the main driving forces. The task that faces design professionals is the management all the variables of strategy, cultural diversity, logistics and delivery alongside ever changing technology. The next section of the book will look at how a number of these elements can be combined in the superpositioning matrix.

3.5 Market positioning, superpositioning and co-positioning

As stated in Chapter 2, market positioning is the act of communicating a distinct position in the marketplace. It is the attempt to distinguish design professionals from other service providers in the minds of existing and potential clients. It is not simply an image making exercise: market positioning must relate to realizable benefits to the client. In other words, clients must know the real difference between design services and the services offered by other professions.

In the case of architects, for example, the profession aspires to represent both the interests of the client, the end user and society as well as taking a lead role in the construction process. In many ways the continuing pivotal role of the architect between client and industry is crucial if the quality of building and the built environment is to be maintained. However, the in-built contradictions and inconsistencies of this market positioning create confusion and misunderstanding in a profession which is struggling to establish a marketing culture. Architects have to compete with a growing range of service providers and have to demonstrate the uniqueness and difference of their service. However, the profession appears unable or unwilling to articulate exactly what the differences are and what are the benefits to the client. It is suggested that the reticence and inability are in part due to the market positioning adopted by the profession. It is hard to market a service when the profession wants to be all things to all people. On what basis are clients expected to demonstrate their preferences for architects? Is the profession sending a consistent and clearly defined positioning message to the market? What is required is a better articulation of the services offered by architects not only in terms of benefits to the client but also in terms of market positioning. Cogent and consistent market positioning is the key to success in the market place for both architects and engineers.

In a scenario plan, market positioning is defined by the interrelationship between the driving forces, predetermined elements and critical uncertainties that shape and form the practice and the target market. Coxe et al. [13] have identified two key driving forces whose interrelationship is felt to provide 'a rational basis for understanding why some firms succeed by doing things one way while others can be equally successful doing things quite differently'. One driving force is the 'process for executing projects and delivering results for clients'. This driving force is termed the design technology. The second driving force is the organizational and operational philosophy of the practice leaders and is given the term 'organizational values'. The two driving forces are then juxtaposed in a 'superpositioning matrix' (see Fig. 3.4).

Organizational values are shown as a continuum between practice led businesses whose primary concern is the delivery of their respective professional service and business centred practices whose main aim is operate a business that happens to be a design consultancy. One of the common instant assumptions is that the business centred practice is commercially successful and the practice centred business is making a financial loss. Not true. Organizational values are not an instant indicator of financial success or failure. The main determinant is how these values relate to the target market.

The vertical axis of design technologies is split into three parts. Strong delivery practices concentrate on the routine execution and delivery of commissions. Strong service practices offer managed services for more complex projects. And strong idea practices that offer creative design, technical innovation and a high level of build quality. The suggestion is that the design ideas and technical innovations of strong ideas

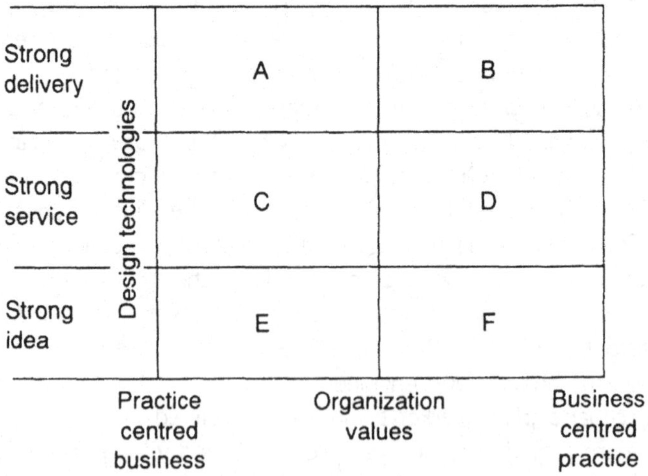

Fig. 3.4 The superpositioning matrix. Source: Coxe et al. [16].

practices are adapted and customized to suit the clients of strong service practices and the most practical and functional elements are eventually incorporated into the work of strong delivery practices. The combined effect of reduced lead times for the design and construction of buildings and improvement information has resulted in a quicker transition from strong ideas to strong delivery. The result is that the middle band of strong service practices is being squeezed between strong ideas and strong delivery practices that are beginning to redefine their position in the marketplace.

Within the matrix, a series of observations are then made about common practice profiles for each of six boxes. The observations come under eight headings:

- project process and decision making:
- organization structure and decision making;
- leadership and management;
- sales messages and type of clients;
- marketing approach and marketing organization;
- pricing and profit;
- leadership and management;
- rewards to the principals.

Suggested master strategies are given each composite type under similar headings. The practice profiles and master strategies that are developed in the book *Success Strategies for Design Professionals* contain a wealth of insight and accurate observation and should be read by every architect and engineer in private practice. This book can only touch upon some of the key observations.

Eric Schneider [26] notes that:

Many but not all of Coxe's findings based on their US research are supported by our own UK observations. For example, 'practice centred strong ideas' firms tend to have creative direction from the partners; their staff tend to be attracted by the reputation of the firm and/or by internal connections; and their clients tend to be or to have key individuals as top level decision-makers or patron-managers. Likewise, larger, 'strong services practices' tend to attract staff more comfortable with a more corporate structure, and often pay more and offer greater benefits. They also tend to allocated the coordination of marketing and new business development under one partner or director and have as their best clients major corporations with large projects, the execution of which the client will generally delegated after making the selection...

Two main weaknesses of the model are: first it assumes that for any one box only one set of factors will suit; and second it confuses a useful device for reflecting internal or peer group perceptions with a mapping or positioning matrix based on client needs and perceptions — i.e. it is not a useful market position device.

Schneider's comments on the inadequacies of superpositioning matrix are valid. Despite the shortcomings, however, the superpositioning matrix represents a quantum leap forward in the way design professionals are able to articulate their situation. In the context of scenario planning, the superpositioning matrix can be a significant building block in the process.

The two driving forces identified in the superpositioning matrix might not be those that you would choose as the most important for your practice. For example, your main concerns might be about the impact of demographic trends (predetermined elements) on your target market or your ability to manage risk (critical uncertainties). You might want to take account of predetermined elements and critical uncertainties as well as driving forces. For example, Hedley Smyth [27] has developed a market position risk profile for flagship urban regeneration projects (see Fig. 3.5). In this case, geographical coverage refers to the extent of the market

Routinized culture	Low	Low-medium	Medium
Service culture	Low-medium	Medium	Medium-high
Innovative culture	Medium	Medium-high	High
	Local	National	International
		Geographical orientation	

Fig. 3.5 Market position risk profile. Source: Hedley Smith [27].

from which investment and consumption will be attracted for an individual building development.

The superpositioning matrix can therefore be adapted to take account of driving forces, pre-determined elements and critical uncertainties as well as other key factors relating to the practice, a sector or an individual project.

Always remember that it is your story. Do not be too concerned about the apparent intricacies of scenario planning and superpositioning. At the end of the day they are only a means to an end. You might, for example, find that your practice positioning varies in several areas of operation. This is common. The variations are likely to be a reflection of how your practice has grown and what opportunities have been pursued. What is important is that these variations are recognized in the scenario planning process and in the subsequent strategic mapping. The scenario plans will tell you where you are now and the options for where you want to be in the future. Strategic mapping tells you what you need to do to get there.

The other important aspect of positioning and superpositioning is co-positioning. Co-positioning is the recognition and resolution of differing market positions. This may be required within a single practice, between two or more individual practices or between different divisions of a larger practice. For example, a large engineering practice operating in several countries might decide that it wants to develop a supportive network of offices that offer a similar service. Staff from each of the offices have a series of meetings, technical aspects are discussed in depth, mutual technologies are agreed and a programme of cooperation and research is outlined. However, when the chance comes to deliver the combined service, the system does not work well. There has been no recognition that some of the offices are positioned differently in the marketplace. The difference in client perceptions and expectations for each office has not been taken into account and result is a mismatch at the point of delivery. The same argument applies to two or more architectural practices that want to network either formally or informally. Each practice must relate its market position to the other practice or practices if the cooperation is to work effectively.

3.6 The Transart story

The year is 2020. Paul Harper has flown into Birmingham airport from a business trip in Prague and his secretary has booked him in to his favourite Thai hotel. Paul had become accustomed to taking off his shoes when entering a room and the heated thick pile carpet offered a reward. He unpacked his bags and took the virtual reality glasses off their wall holder. Experienced fingers selected 'Historical Events', '1966 World Cup', 'Terraces'. He had decided to view the match from the standing terraces this time. It would give him a chance to jump up and down when England scored. It was the nearest he came to aerobic exercise in his busy schedule. Besides, he had been sat at a desk or in a plane all day.

Paul knew the early stages of match by heart and his mind wandered to his days at the Rotterdam Academy. He had been a mature student and had graduated in 2003. His eight-semester course in architecture, engineering and fine art had been one of the first of its type in Europe. Two of his friends had graduated at the same time and they formed a small company specializing in transport engineering and urban art. By 2014, the company had grown to twenty people and they had built up a reputation in the field of transport interchange design. The European urban planning regulations passed in 2005 had restricted vehicular access to major cities and Paul's company, Transart, had responded to the opportunity.

In 2008, the Japanese company, Sibushi, invented free-form holograms. A complex optical diffraction system made holograms look solid and real. They launched the office model as a one-metre square cube projection that allowed designers and clients to get an all round view of any proposed building. Additional software provided, for example, interior views, walk throughs and outline costs. The capital cost of free-form equipment was prohibitively expensive. Sibushi struck a deal with a French company, Artel, which acted as a European concessionaire. By 2013 Artel franchises had enrolled over a third of all architects and design engineers in Europe. Employment was spasmodic but the pay was good.

The videophone signal interrupted Paul's viewing. It was Hans Schmidt, Paul's contact with the Ipkros Group. Paul had never understood fully how Ipkros operated. All that he knew was that the sharp recession in 2002 and the later banking collapses had led to calls for market control and regulation. Paul knew that the Ipkros Group conformed to the European market rules and regulations applied in 2009. Ipkros also operated as a work finder for Transart: a marketing service that had become increasingly important for Paul's company. The transport inter-change market in Europe was in decline. Paul had to travel further and further afield to find suitable commissions as well as diversifying into other types of work. Transart employed now only eight staff. Most of the basic work was given to people from the local Artel franchise employed on short-term contracts. Paul felt confident that Transart's designs remained confidential following the implementation of the European Data Protection Act in 2008.

When Paul was first approached by the Ipkros Group he had tried to find out more about the organization. A company search had shown that two of the directors on the main Ipkros board were also directors of a large German construction company. Paul was unconcerned about this finding because he had been given complete design freedom in five years working with Ipkros and there appeared to be no pattern of contractor selection by Ipkros for each job.

Hans was smiling on the screen and his affability oozed through even the videophone. 'We have a possible job in Bremen, Paul. I've spoken to the Risk Management Department. They appear to have split the funding risk between a Swiss Bank, a Taiwanese fund holder and a Korean pension fund. They are all Euro rated but the competition is fierce. We have ten days to make an outline design proposal.' In total there were five pan-European syndicates that competed in certain markets. Traditional developers had long ceased to exist in the market as their Euro risk management rating had not been high enough. That reminded Paul. Transart had not sent in a copy of this quarter's accounts to the European Practice Registration Board. The Board had existed since 2005 and in the last few years, the information Transart supplied to the Board was also used to determine Transart's Euro rating.

In 2013, Transart had won the Hamburg Architecture Centre's gold award for the design of a transport interchange on the outskirts of the city. The design of the interchange had been an open competition. Paul had been approached subsequently by three syndicates and he had eventually chosen Ipkros. There was no formal agreement, a situation that suited Paul. Working with Ipkros had the additional bonus of reduced rates with Artel and priority treatment on Ipkros jobs. The latter arrangement had been the deciding factor for Paul. He hated negotiating with Artel and having to ask them for enough good designers and technicians to do the work. Less hassle would give him more time to concentrate on design issues.

Hans had been Paul's contact since the very beginning of the relationship with Ipkros. At the start, Ipkros and Transart had been successful in winning commissions for a small number of high profile designs. In the first year of cooperation Transart had started work on eight commissions but in ensuing years the number of jobs had increased considerably. In the last year Transart had a rolling programme of over fifty jobs and many of the designs had a sameness that concerned Paul greatly. About two years ago, Paul had dinner with Hans at a restaurant in Brussels. He complained that his team did not have enough time to work on new approaches. Why were they having to work on such a wide variety of projects? Couldn't Ipkros find fewer and better projects for Transart? Hans remained non-committal and suggested that they took a stroll in the old quarter of the city. The Belgian beer was good and Hans talked about vertical integration of contractors, sliding down the superpositioning matrix, the need for strategic mapping, programming for innovation, technological dominance, and maintaining market leadership. Paul did not understand a word that he was saying and wished that he had taken more notice of the marketing and business development lectures at the Rotterdam Academy. The effect of the Belgian beer did little to help his thought processes.

'Hans, I'm tired. I've just flown back from the site in Prague. There are problems with the contractor. Can your people talk to the contractor about the detailed drawings and the material specifications? I will need time to work on a submission.'

'No problem Paul. I'll get on to our Prague office in the morning. Perhaps you can ring me on your car phone later in the morning to discuss the new project.'

'OK Paul. I'll give you a ring. I think that we must have a talk again about where we are going.' Hans smiled, said nothing and waved goodbye. Paul picked up the virtual reality glasses and selected the Royal Box this time.

3.7 The Transart story: comments

The Transart story is neither a prediction of the future nor a scenario plan. It is a fanciful tale that has been made up to take the reader outside the current preoccupations of the 1990s and to illustrate a number of points about the scenario planning process.

Imagine that you are a consultant who has been asked to advise Paul shortly after winning the Hamburg Architecture Centre award in 2013. Your appointment has been the result of a discussion between Paul and his former tutor at the Rotterdam Academy. In the story, Paul's tendency to hark back to former glories did not help his situation in 2020, and similarly his singular concern with design is equally disempowering in 2013.

All Paul wants to be is a good designer. He has little interest in the outside world. His talents have been recognized internationally and he considers his professional status and skills inviolable. His view of the world is such that his relationships with Ipkros and Artel pose no threats. Indeed, the relationships appear to offer more freedom. He has more time to spend doing design work and the burdens of managing staff and finding work fall on other shoulders. In my opinion, the opposite is true. His attention to the operational, business development and marketing aspects of the company are the very things that empower and lead to success in the future. If Paul had spent only a relatively small amount of time looking at the social, economic, political, legislative, cultural and technological developments in a few scenario plans in 2013, he would have been much better placed in 2020.

The specialist skills that made Paul's company so successful at the beginning were ultimately a limiting factor in the long-term development and sustainability of the organization. In the next chapter there will be a detailed look at how design professionals can counter the threat of specialization through a process of synthesis marketing. Synthesis marketing is a process of bringing together a wide variety of project elements in a number of selected markets.

Synthesis
marketing

4

The main conclusion to be drawn from Phase 2 of the RIBA's 'Strategic Study' is that the present depressing circumstances must be seen as creating excellent opportunities for architects. But this will happen only if architects are prepared to devote as much design imagination to managing their relations with clients as they devote to crafting their client's buildings. (Frank Duffy, former RIBA President [28])

4.1 Synthesis marketing explained

Marketing in the twenty-first century will be shaped and formed by synthesis rather than analysis. Successful marketing strategies in the next century will be based on:

- a synthesis of social, cultural, political, economic and environmental factors;
- a demonstrable ability to bring together and manage a wide variety of project elements;
- a clear articulation of the benefits of intangibles such as design, quality and purpose.

In many ways, the theory and practice of marketing were hijacked by the economic imperatives of consumer goods industries in the 1950s and 1960s. The rubric of traditional marketing developed during this period is based on quantitative retrospective analysis used to predict future trends. This approach is inappropriate for emerging markets involving new technologies and new methods. How can one predict the size and nature of the environmental market for design professionals using historical data? Any success in this field will only be won by establishing a market position in selected markets with distinctive roles and matching competences. Individual organizations and their respective professions will have to face up to the intuitive, forward looking challenges that this presents.

If design professionals are to abandon the analytical tools of traditional marketing, what will replace them? What will be the building blocks of a new synthesis marketing approach? Figure 4.1 provides a diagrammatic representation of a way forward.

Marketing valence is a measure of the power of attraction between your organization, the client and the sectoral infrastructure. The concept raises many questions. What binds you to this group at this point in time? How is your attractiveness to this group likely to change in the future? Are your chosen marketing strategies the best way to bond or link with this group? What is your intuition about how the marketing valence will change in the future?

Marketing composition is a check that you have the right ingredients for success in the chosen sector. Are you providing appropriate services to satisfy client requirements? Do you need to join forces with other service providers to meet existing and future client demands? Are your services adaptable enough to meet changes in marketing valence?

Entelechy is a word coined by Aristotle. It was used by Aristotle in a

Fig. 4.1 The synthesis marketing building blocks

philosophical sense to mean the transition from potentiality to actuality. The complexity of the situation can vary. For example, clay becomes a brick or the universe unfolds in an infinitely complex way – both through the process of entelechy. In the rather more mundane world of modern business, the concept of marketing entelechy springs from a belief that given a specific mix of players, marketing composition and marketing valence, a particular outcome is likely. Potentiality becomes actuality through the process of marketing entelechy.

The synthesis factors shown in Fig. 4.2 are those which make your organization attractive to the client and the sectoral infrastructure. The synthesis factors appropriate to your company will not necessarily be those shown in the diagram. You are free to choose your own set of factors but your choice should be informed by feedback from clients and the related sectoral infrastructure. Synthesis factors are the factors that determine marketing valence, reflect marketing composition and enable marketing entelechy. The importance and contribution of each synthesis factor is reflected in its position in the synthesis marketing matrix.

Marketing valence, composition and entelechy combined with synthesis factors are the building blocks of a new marketing discipline for architects and engineers. However, before the concepts are explained in more detail, it is necessary to look at what is represented by the transition from traditional to synthesis marketing. It is assumed that the architect or engineer reading this book will have only a slight understanding of traditional marketing methods. A greater understanding of traditional

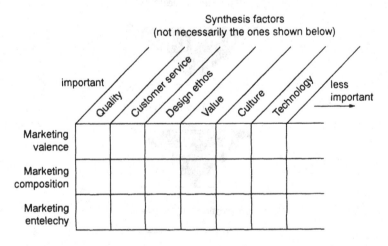

Fig. 4.2 The synthesis marketing matrix

marketing methods is not however required. What is needed is an understanding of the nature of the transition from traditional to synthesis marketing and the new direction of synthesis marketing effort.

4.2 The transition from traditional to synthesis marketing

We have moved into a new era in which physical products are viewed as part of a wider service and service industries, including the professions, are being packaged and sold as products. The only way to resist the commoditization and packaging of professional services is to redefine the relationship with the client, the marketplace and society. Traditional marketing methods are inappropriate and inadequate to deal with the complex market positioning of design professionals in a modern economy.

Traditional marketing uses the concept of market share, i.e. the proportion of work won in a particular sector, as a measure of performance. Few architects or engineers have a significant proportion of work within in any particular sector, within, say, the European economic region. In the future, architects and engineers will have to address markets in which their measure of performance will be their ability to sustain their relationship with both individual clients and the corresponding sectoral infrastructure.

Design professionals will have to gain market acceptance in individual sectors through their synthesis marketing efforts rather than across the board market awareness gained through traditional promotional measures. Architects and engineers have to be seen to be serious players in any market. Architects in particular are faced with a long-term struggle to win and retain market acceptance and understanding of their services.

The traditional formalization of marketing effort is being replaced by flexible marketing effort. The latter draws upon the efforts of everyone within an organization rather than just individuals or senior members of staff.

Traditional marketing assumes that market positions are fixed. Synthesis marketing allows positioning to be adapted to suit individual markets. Design professionals will have to balance consistent positioning, which is likely to lead to success, with the requirement to respond to the needs of the changing marketplace. The strategic mapping methods discussed in Chapter 5 must take into account possible changes of positioning throughout the period of the scenario plan.

4.3 Synthesis and relationship marketing

The assumption made by the traditional marketing approach is that there is a point in time when a transaction occurs. This type of marketing is sometimes referred to as transaction marketing [4]. In the service sector there is also a point in time when, for example, a commission is made, but the relationship with the client can span several years. A marketing strategy geared solely to a single point in time would therefore be totally inadequate and inappropriate. It is not unusual for an architect to meet a potential client at say an exhibition or conference and two years later for a contract for a commission to be signed. The design and construction of the building could then take another year and a second commission for another building might result several years later. How do you formulate a simple, systematic and sustainable marketing strategy to cover such a long period?

From the mid-1980s, there has been a move away from traditional, transaction marketing approach towards a longer-term, enduring relationship with the client and the sectoral infrastructure. This has been called relationship marketing. The primary concern of relationship marketing is the linkage of quality, customer service and marketing (see Fig. 4.3). 'Traditional marketing has been about getting customers. Relationship marketing addresses the twin concerns – getting and keeping customers' (Christopher et al. [29]).

Martin et al. [29] emphasize three issues:

Fig. 4.3 The relationship marketing orientation: bringing together customer service, quality and marketing. Source: Martin et al. [29].

- 'Relationship marketing strategies are concerned with a broader scope of external market relationships which include suppliers, business referral and 'influence sources'.' In this book, Regis McKenna's term 'sectoral infrastructure' has been used to describe these external relationships.
- 'Relationship marketing also focuses on the internal (staff) relationships critical to the success of (external) marketing plans. 'Internal marketing' aims to achieve continuous improvement in marketing performance.'
- 'Improving marketing performance ultimately requires a resolution (or realignment) of the competing interests of customer, staff, and shareholders, by changing the way managers manage the activities of the business.'

Relationship marketing therefore represents:

- a move away from a single point in time to a more enduring relationship with the client and the sectoral infrastructure;
- a recognition that internal marketing and management must relate to the external marketing plans

From the point of view of architects and engineers, relationship marketing has been a necessary big step away from traditional marketing and many of the points that are raised are helpful and valid. However, there is a considerable assumption that customer service and quality are the two deciding synthesis factors in any relationship with the market. For design professionals this is unlikely to be the case in many markets and other synthesis factors will be more important. Relationship marketing needs to take the next big step towards synthesis marketing with its wider definition of marketing valence.

4.4 Synthesis marketing and cultural consumption

As stated in Chapter 3, cultural consumption and cultural consumers are firmly established in a modern economy. According to Stuart Hall [23]:

> If Post-Fordism exists then its is as much a description of cultural as of economic change. Indeed, that distinction is now quite useless. Culture has ceased to be, if it ever was, a decorative addendum to the 'hard world' of production and things, the icing on the cake of the material world.

The aestheticization of consumption has become the consumption of aesthetics. Cultural consumption and consumers can range from the design of a power station or a factory to an individual who wants to buy a piece of furniture for his/her home. The distinctions between culture and consumption no longer exist and design professionals are having to respond to the marketing challenge. Design professionals will have to learn how to deal with the market's demands through a synthesis marketing approach that has a cultural dimension as one of the important synthesis factors.

The key marketing question for any industry or profession is 'what market are we in?' It is often better to identify a wider generic service market in order to understand how a number of related services can be brought together. The wider market opportunity needs to be identified so that the complete picture of the real needs and wants of the consumer are fully understood. For the purposes of synthesis marketing for architects and engineers it is assumed that:

Design professionals are in the cultural consumption market.

It is perhaps difficult to understand that an engineer designing a bridge is offering a cultural as well as a technical service. It should be recognized that the firm of engineers designing the bridge has its own cultural identity and design culture and is working for a client who is consuming the collective culture of the service organization. Marketing is no longer an issue of demonstrating technical competence and delivery capability to a client's satisfaction. In the 1990s and in the next century, the client will want to know a lot more about the design, business and organizational culture of the supply company.

In earlier chapters reference was made to the the existence of unsophisticated clients, many of whom are the product of a devolution process. However, in contrast, large engineering companies have to deal with sophisticated clients who are able not only to negotiate, monitor and control a commission in a far more rigorous way, but are also able to express a cultural preference. It is the latter cultural preference that is causing so many of the existing marketing problems for larger engineering companies. There is, in some cases, an internal cultural mismatch between the various departments and sections of large engineering companies and a corresponding external mismatch with the client and the sectoral infra-structure. A common symptom of the mismatch is a feeling that there is poor communication between the various groups. The engineer's response is often an improvement to the hardware of the communication system, i.e. a technical solution to an apparently technical problem. However, what is needed is a cultural solution to a cultural problem.

Synthesis marketing offers a way of addressing the cultural consumption market and its related cultural problems. It places importance on the qualitative aspects of marketing valence, ensures that the necessary ingredients are present in marketing composition to give the desired result and recognizes the creative process of entelechy that only design professionals can deliver.

4.5 Cultural positioning

The picture that has emerged of the change from traditional to synthesis marketing is a move towards the marketing of qualitative, intangible factors such as design, quality and service. In the case of architects and engineers, it is possible to envisage a cultural dimension being added to the list of intangibles. Culture is the overarching theme in the complex positioning of design professionals with clients, society, end users and the construction industry. Culture, however defined, has to be one of the important synthesis factors if cultural positioning is to be achieved in a cultural consumption market. Cultural positioning assumes that there is a cultural best fit between the company, the client and the sectoral infrastructure. Identifying and producing that best fit is the marketing task of cultural positioning.

In Chapter 2 it was suggested that architects' ambivalence to membership of the construction industry and the resulting mixed positioning message are a liability. Indeed they are. However, the mixed positioning can be made into an asset if the application of synthesis marketing methods result in cultural positioning which will influence cultural consumption, support client and sectoral retention, and assist design professionals in new market entry. It is a complex solution to a complex backdrop. The idea of cultural positioning influencing cultural consumption is part of the rationale behind the 'culture, community, construction' marketing platform adopted by the proposed new Northern Architecture Centre (see Chapter 7). Cultural positioning and its influence on cultural consumption is seen as the first important step in the process of synthesis marketing for both architects and engineers.

4.6 The decision making unit (DMU)

There is a tendency among architects and engineers to embody a single person in the client hierarchy with the title 'client'. He or she is suddenly bestowed with the god-like qualities that befit only a paying client. This tendency to concentrate the marketing effort on an individual is an understandable human failing but is not one to be encouraged in a marketer, marketing partner or director. It is a serious marketing failing to focus attention on the individual concerned at either the initial contact stage, when winning the work, or sustaining marketing effort during or after the commission.

The skill that has to be developed is the art of identifying, contacting and influencing the decision making unit (DMU). The DMU is the group of people who contribute to or have significant influence on the decision making process. The marketing task is to gain sufficient information to understand the workings of the DMU. For example, who is the leader of the group? Who makes contact with the outside world? Who is responsible for bringing together the related issues and information? Who monitors the funding? How can you gain access to these individuals either collectively or individually?

It is a common experience to have the first client contact with a member of the DMU who is relatively low in its hierarchy and does not necessarily have a complete picture of the project or understand all the related issues. There is no magic formula to widening the level of contact or increasing your understanding of the issues surrounding the project. At each stage you have to 'qualify' the individuals that you meet. This means that you have to determine if they are part of the DMU and to come to a decision as early as possible about whether or not the project comes within the preferred portfolio outlined in your marketing strategy. Qualifying potential clients and their projects is a critical skill if you are not to waste a great deal of time and energy. Marketing to a universe of opportunity with an ill-defined marketing strategy is a bottomless pit of work, is incredibly time wasting and is only likely to bring on your first heart attack.

In the synthesis marketing process, you are trying to market a cultural consumption service to a group which has its own collective culture. This is a point that can be easily missed if you put too much emphasis on contact with an individual. In this situation, what marketing valence and synthesis factors will allow the two groups to bond and be held together for the period of the commission? Is your perception of the likely outcomes of marketing entelechy the same as the DMU's? Is your proposed marketing composition for this project understood and accepted by the all the members of the DMU group? Trying to understand these issues at the early stages can help you to resolve problems later on.

4.7 Communication and influence

Many people believe that marketing is synonymous with communication. However, marketing could be more accurately described as being synonymous with influence. Communication is the process, influence is the outcome. Why concentrate too much on the process when it is the outcome that is important? In synthesis marketing, design professionals are faced with the task of marketing to existing and potential customers as well as the sectoral infrastructure. The implication is therefore that architects and engineers will have to establish wider influence. Here again, the mixed positioning of design professionals can be an asset rather than a liability. Architects in particular are used to taking into account the needs of several audiences, including the client. What is sometimes lacking is the feeling that the customer comes first. The only way to resolve this situation is to develop a better understanding of the nature of influence at all levels. Design professionals will have to exert influence in a new environment, an environment of 'choice, options, information, selectivity and access' [2] (McKenna).

4.8 Synthesis marketing management

Throughout the process of synthesis marketing the management effort should concentrate on the client's expectations, perceptions and experience of the service (see Fig. 4.4). You will need to draw up a list of

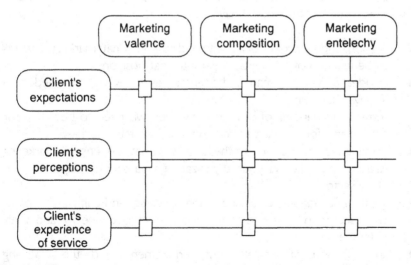

Fig. 4.4 The synthesis management matrix

appropriate questions for each commission. Basic questions can include, for example:

- Are the skills and competences that are being offered in the marketing composition still those required by the client?
- What is the current state of our bond with the client and how is the marketing valence changing?
- Is your vision of the overall process still the same as the client's? Review marketing entelechy.

It is suggested that the checklist of questions should be written down. You will only get the right answers if you ask the right questions. In many cases they are questions that you should be asking as a matter of course. The only difference is that you now have a better structure. The benefit of introducing the discipline is more apparent when dealing with complicated forms of procurement such as novation. In such a case your attention to marketing valence, composition and entelechy is an important part of the process.

4.9   The synthesis marketing process

4.9.1 The strategic marketing plan

It is assumed at the start of the process (see Fig. 4.5) that your practice has a strategic marketing plan that has been developed using the scenario planning, synthesis marketing and strategic mapping methods outlined in this book. Your practice should have:

- a good understanding of its future direction, what markets it wants to be in, and how it wants to get into that position;
- an idea of what human and financial resources will be needed to achieve these aims;
- some understanding of how proactive you will need to be with your marketing effort in each sector and what are the priorities;
- some indication of the synthesis factors that will ensure marketing valence and establish your organization as a serious player in your niche sector;
- some understanding of the common processes and culture of a particular sector in order to determine marketing entelechy and likely outcomes;
- an indication of typical client requirements and the matching marketing composition that will lead to a commission.

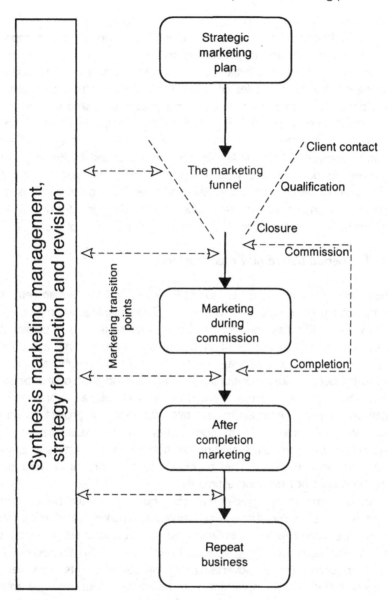

Fig. 4.5 The synthesis marketing process

The initial marketing strategy has to be detailed enough to ensure that, for example, when contact is made with a prospective client, the opportunity can be quickly assessed. Is the project within the preferred building types or market sectors? Is the work within the capacity of your organization?

How would this project fit within the overall portfolio of your company and what further opportunities is it likely to offer?

Bear in mind that the initial marketing plan will be subject to continuous change. It is not a tablet of stone. Limited research of and contact with a particular market sector can reveal previously unknown opportunities and problems. Fine tuning or even a complete rethink of your approach may be required at almost any point in the process. These conscious changes to the marketing plan are marketing transition points. They need to be built into your system in order to assess the client's current state of perception, expectation and service experience. This is a qualitative assessment that in many cases will be done intuitively; but it is worth including marketing transition points in your pattern of work.

4.9.2 The period before new client contact

In their book, *In Search of Excellence* [30], Peters and Waterman refer to the importance of a 'volunteer champion'. The volunteer champion is a person who is self-motivated, does not need cajoling and has a passionate and sometimes irrational interest in the subject area. Finding and developing these people is a management task in its own right and is an important aspect of superpositioning, i.e. the development of the people market. The volunteer champion could be, for example, a junior member of staff who has the enthusiasm and talent to take the project forward. He or she may make a vital contribution to the reassessment and reappraisal of the company's existing client base or may act as a springboard into new markets. The potential contribution of a volunteer champion should not be underestimated.

Another important ingredient which should be initiated before client contact is a set of measures which promote market acceptance. The difference between market acceptance and market awareness is worthy of note. The traditional marketing approach has been to raise awareness of a product or service through advertising or promotional events. However, it has become increasingly apparent that awareness does not guarantee acceptance. A prospective client might be aware that your practice has done some work in a particular sector but may not necessarily accept your firm as a serious player in this field. The alternative approach is to readdress the sectoral infrastructure. As stated previously, the sectoral infrastructure is defined as every organization and individual that can influence client perception. The infrastructure of any sector may include trade organizations, trade press, client associations, sectoral institutions, governing bodies or consumer groups. For small practices operating within a regional economy, the infrastructure may also include key organizations and institutions that make up part of the local economic network. It is unlikely that advertising and press releases alone will be sufficient to ensure acceptance or indeed be relevant to your particular sectoral infrastructure. What is needed is a clear understanding within your company of the message that you want to communicate combined with a programme of contact with members of the sectoral infrastructure. Time spent in contact with organizations and individuals within the sectoral infrastructure is never wasted.

4.9.3 The marketing funnel

The marketing funnel covers the period in the process from initial contact with a potential client to securing a commission. It is divided into three parts:

- initiating and managing client contact;
- 'qualifying' the client;
- 'closing', i.e. winning the commission.

The marketing funnel is the part of the process usually associated with marketing. The management task is to reduce a large number of possibilities down to a handful of commissions. Ideally the commissions that you eventually win will not only provide work but will also help you to move towards the longer-term goals of your marketing plan. At a time when information about both public and private sector commissions is more readily available, the management of the marketing funnel becomes an important part of the process.

Initial contact with a potential client can be, for example, in the form of a response to an advertisement or a face to face meeting. The quality of the initial contact can vary considerably. If you meet a prospective client at, say, an exhibition or conference, the exchange can leave you with more questions than answers. Is funding available for the project the client has in mind, and what is the timescale for development? Are other professional services, such as project managers, involved already? What form of procurement is preferred by the client? You need to build up a picture of the prospective client and project quite quickly if you want to be sure that further contact will be worth the effort.

The first marketing transition point is often a record of your changes of perception after initial contact with a new sector or immediately after contact with a potential client in an existing sector. If you are, for example, entering a new market it is often the case that the adjustments to your marketing strategy are greater than expected in the early stages. Limited background research cannot adequately cover the dynamics of even the simplest markets and there is no substitute for direct contact with the sectoral infrastructure and potential clients.

The aim of the first contact should be to establish confidence and trust with the client and to start what you hope will be an enduring and mutually beneficial relationship. A graph showing a client's confidence curve is shown above in Fig. 4.6. The shape of the curve will vary considerably with every job. To win a commission, your aim as a marketer should be to raise the confidence and trust level as high as possible before a contract is awarded.

During the period before a contract is awarded you have a number of aims such as addressing the design issues that face your client, making all members of the DMU aware of the range and quality of the service that is on offer, and managing the contact. Bonding with a client, i.e. the DMU, is a difficult and complex process. The tendency to offer repeat business is based on a confidence that your company can problem solve and that you have produced satisfactory results. In the case of a prospective client you have to instil that trust and confidence, sometimes in a short space of time.

All prospective clients have a hidden agenda. This may be a series of issues and concerns which they want addressed but may not be able to articulate in a way that can be translated into a building or plan. Your first task as a marketer is simply to listen. You must reflect positively on what is being heard and learn the language of the client. Professor Randolph

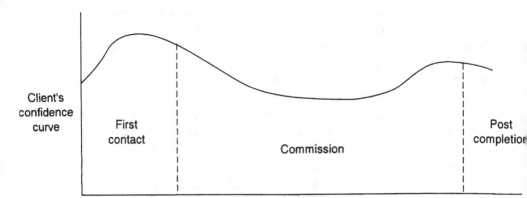

Fig. 4.6 The client's confidence curve

Quirk [31] wrote that 'every particular use of English is to some extent reflected in and determines the form of the language that is used for that particular purpose'. The form of the language used by, say, a health services manager can be dramatically different from that spoken by an architect new to the health sector.

At an early stage, you have to come to some preliminary agreement with the client about the quality of design, the required level of build quality and the amount of technical innovation. From the design professional's point of view, these issues are very important. From the client's position, they may be far less important and may be not be appreciated. It is very easy for the architect or engineer to get bogged down by technical detail and to dwell on the finer points of previous work. The client's interests are more likely to be in the quality of service that is on offer. For example, will the project be on time and within budget? What construction supervision systems are in operation? Does the practice have a total quality management programme? To a certain extent, this form of checklist procurement has been taken too far in the first half of the 1990s. The changes in procurement patterns and the opening up of the public sector have resulted in clients drawing up long checklists of management and operational criteria. In many cases these checklists are more about meeting statutory obligations than meeting the real needs and aspirations of the client. At the same time, the process places an undue emphasis on tender prices. If your company and several others have managed to get ticks in each of the checklist boxes, the only remaining criteria are cost and the ill-defined aspect of good design. The only way to counteract this process is to define good design in a way which has meaning for the client. For example, good design for the client may mean lower running costs and maintenance.

Your approach to resolving the balance between service and design will depend on your superpositioning organizational values (see Chapter 3). Do you want to have a practice centred business or a business centred practice? A practice centred business will have to decide if the commission gives it the required qualitative rewards and feelings about their work. A business centred practice will place more emphasis on the quantitative rewards. The end result, in terms of the level of service and design may, in some instances, be the same for both types of practice. In these cases, only the practice perceptions would be different.

The prospective client is faced with the task of making a choice. The task is in two parts. First, to decide on what grounds the choice will be made, and second to make comparisons between what is on offer from each firm. Your task as a marketer is limited strictly to influencing the method and nature of the choice. Never draw comparisons with competitors and never say anything bad about another company. Criticism of competitors will always reflect badly on your organization. However, that does not mean that you cannot place competitors in an unfavourable position. For example, if one of your competitors has a run down office with few computers and you have a new office and a new CAD system, suggest to the client that a visit to the practices is included in the selection process. Be knowledgeable about your competitors but never draw comparisons in front of a client.

4.9.4 Marketing during the commission

Synthesis marketing involves not only a synthesis of service provider, client and sectoral infrastructure but also a bringing together and management of a wide variety of project elements. It is important therefore that the cultural positioning and marketing valence that has been established pre-contract continues throughout the commission. There is no point in putting together all the appropriate ingredients in the early stages if there is a subsequent imbalance in service delivery and a move away from what was agreed.

There is a common perception that at the commission stage the marketing effort is complete and the job is won. However, an equal amount of marketing effort should be put in during and after the commission to ensure that the promised services are delivered and that the changes required to marketing valence, marketing composition and cultural positioning are understood by all parties involved in the process.

4.9.5 After completion marketing and repeat business

Research by the Royal Incorporation of Architects in Scotland [32] has shown that in Scotland

> an average of 49% of all work in a given year is for a client for whom the architect has already worked. It reduced to 20% for under one-fifth of the firms, whilst one-third of the offices enjoyed 70% or more; and for two practices (in the survey) it was distinctly unusual and very rare to work for anybody else.

Anecdotal evidence suggests that the above figures may exaggerate the level of repeat business but they do illustrate the point that a significant proportion of core business is likely to come from previous clients or from prospective clients with whom there has been contact over a long period of time. If the work comes from a sustained period of negotiation with either an existing client or a prospective client, the sooner the process of negotiation can be started the better. In the case of the previous client, the marketing task after completion of the commission is to maintain contact and to monitor the changes in the client organization. The latter effort is the reason for having a marketing transition point between project completion and repeat business. Do not assume that the marketing valence of the previous commission will be the same for the next project.

4.10 The cost of synthesis marketing

How much human, financial and technical resources should be allocated to synthesis marketing effort? There is no rule of thumb about the recommended level of resources needed to win or retain work. Misleading figures are often quoted about the proportion of overall turnover needed to ensure adequate marketing effort in this area. For example, an often quoted figure is that, in the 1990s, architects have had to increase their spend on marketing from 2% to 4% of their overall turnover. These comments do not stand up to much scrutiny. Often, the calculations do not accurately take into account the time spent by both junior and senior members of staff on general marketing effort. There is often a reluctance to face up to the fact that marketing, in all its aspects, accounts for a much larger proportion of overall turnover than directors or partners are willing to admit. Other service industries have recognized that their marketing effort can account for up to 30% of their turnover. These organizations regard marketing as part of the business process and not just as an overhead that is an adjunct to the 'real' work of service delivery.

There is no set figure or formula for the cost of synthesis marketing for architects and engineers. Costs will vary considerably between sectors and size of practice. However, the point to note is that if synthesis marketing is to be implemented effectively, full account should be taken of marketing costs. There is nothing to be gained by ducking the issue of human, financial and technical resources. Synthesis marketing should be an integral part of the overall service and as such requires adequate resources.

5

Strategic mapping

In modern deregulated markets, you either attempt to guide your business destiny, or let destiny have its way and learn through the pain. (Martin et al. [29])

5.1 What is strategic mapping?

Physical maps redefine our understanding of physical reality. Mental maps redefine our organization of conceptual reality. They are the way in which we try to impose some order on a world of apparent chaos. These mental maps embody all the hopes, fears and aspirations that colour and shape our mental construct. They identify what we feel is of value and what we think is needed to achieve success.

Strategic mapping identifies strategic options available to practices and then maps them out over the time period of each scenario plan. The strategic emphasis is on choice and empowerment. It is only when you have both the scenario plan details and the related strategic maps that you can select a preferred scenario plan. You need to have as much information as possible about the long-term view and the stages in between before you can make a choice.

So why bother to look at strategic options so far in advance of events? The first reason is straightforward, namely to benefit from facing up to possible problems in advance and being able to work on solutions without the pressure of events. The second is more subtle and requires some explanation. Marketers and people involved in marketing suffer more than most business professionals from what is sometimes called the

'*Einstellung* effect' (Luchins [33]). The theory behind the *Einstellung* effect is that when a person or group of people find a simple solution to a particular problem, they will tend to find a more complicated solution to a similar problem in the future.

In the case of marketing, this is often apparent in, for example, new market entry problems. A company decides that it wants to make inroads into a new market. It allocates sufficient resources, contacts key decision makers in the target sector and conveys a distinct message to the sectoral infrastructure. Several years later, the company decides to enter a new niche market. This time the problem appears to be more complicated. More resources are required, the apparent problem is more complex, more preliminary market research is needed and everyone is less confident, no matter what the outcome of the previous market entry.

It is suggested therefore that long-term strategic mapping offers a way of forming a view at a point in time and in a situation uncluttered by the *Einstellung* effect. Good marketing, like good design, is seen at its best in simple solutions.

5.2 Pre-emptive marketing

The development and refinement of pre-emptive marketing is the starting point for strategic mapping. According to Pickar [34]:

> *Survival and success will depend on a company's ability to adapt trends in the marketplace and to gain solid, current information from narrowly focused markets. Successful firms will also engage in good, flexible, strategic thinking about what they should be doing and what marketing strategies will meet these goals ... as the marketing gospel has spread, more firms have moved from reactive marketing (waiting for a project announcement) to proactive marketing (approaching the client before a project is announced). In the '90s progressive firms will move further, to pre-emptive marketing (offering solutions to the earliest problems a potential client may have).*

During the last recession, it would have been possible, for example, to have anticipated three periods of market change: a period of survival, a time for consolidation and an opportunity for later expansion as confidence returned. Although potential clients might not have been in a position to commission work during the recession, they had an opportunity to have a strategic rethink about their plans for the future. At the onset of the recession it was possible to envisage a strategic map along these lines. You would not have known how long the recession would have lasted or how deep it would have been. However, it was possible to look at a set of scenarios that took account of the survival, consolidation and expansion phases. In the event, this would have been a scenario plan of almost six-year duration.

The development of the scenario plan after the beginning of the recession would not have required any great intellectual prowess or knowledge of global economics. It would have taken time to think about what was likely to happen and an ability to put oneself in the shoes of potential clients. The same logic can be applied to the remainder of the 1990s. There will undoubtedly be another recession. History suggests that there is one about every ten years. It is possible to envisage a period of expansion and confidence with a varying rate of growth in the market. When the next recession comes its onset will be as quick if not quicker than the previous recession because the financial markets can easily turn off the tap of investment and market confidence is correspondingly fragile. What position do you want your company to be in by the next recession? What changes will you make in your marketing and organization to survive in the long term and flourish during the good times? Taking a ten-year view does not seem unimaginable if you look at the market in this way. You must have a 'willing suspension of disbelief' in order to develop a more intuitive response and a forward looking, pre-emptive marketing stance.

Strategic mapping, as part of a scenario planning and synthesis marketing exercise, is the route to success at both corporate and professional levels. Design professionals must begin to anticipate and influence changes in their environment. The challenges which are likely to face design practices in any strategic mapping exercise will include:

■ the integration of a synthesis marketing function into the operation of your practice;

■ the need to manage and organize your company so that it is able to respond quickly and effectively to shifts in market positions. Organizational forms will have to be changed to suit future markets;

■ an acknowledgement of interdependence in whatever target market is identified. Your company will need to develop strategic alliances with other organizations if it wants to survive the competitive environment of the future;

■ a concurrence of cultural, organizational, technological and marketing effort to meet individual challenges;

- a better understanding of how you can influence future events, and in particular the strategic thinking of potential clients;
- an appreciation of the impact of information technology as an integrating factor between various parts of your organization, your strategic alliances, and your client base.

5.3 The four marketing choices

Practices are faced with four marketing choices:

- marketing existing services to existing clients;
- marketing existing services to new clients;
- marketing new services to existing clients;
- marketing new services to new clients.

These four strategic options are sometimes represented diagrammatically in Ansoff's matrix [35], shown in Fig. 5.1.

The Ansoff matrix offers a systematic way of exploring the basic strategic options. The option with the least risk is to continue to offer the same service to the existing client base. The strategic option with the highest risk is to offer a new service to a new market. Making only one strategic change of either moving into a new market with an existing service or offering a new service to an existing market has a risk level somewhere between the two extremes.

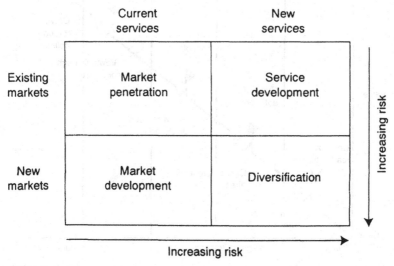

Fig. 5.1 Ansoff's matrix

A simple way of looking at strategic options within Ansoff's matrix is to do a SWOT analysis. The strengths and weaknesses of your practice are related to the opportunities and threats in the marketplace. SWOT analysis gives a view of both the internal and external constraints and provides some indication of the level of risk and the strategic problems that are faced.

5.4 The practice growth cycle

Ansoff's matrix combined with SWOT analysis is a crude but effective first stage in the strategic process. It can however offer a better insight into the present and future prospects of the company if the results are viewed in the light of the position of the organization in a growth cycle. Greiner's model [36] (see Fig. 5.2) is a useful five phase growth model which is particularly relevant to professional service organizations. The five phases are characterised by both evolutionary and revolutionary stages. The term evolution describes 'prolonged periods of growth where no major

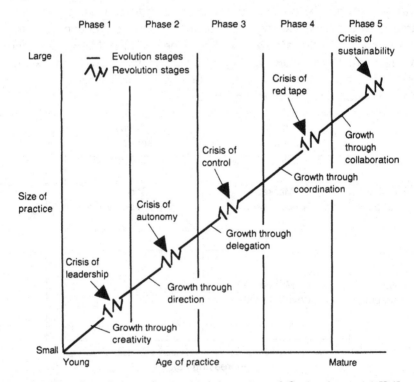

Fig. 5.2 The five phases of growth. Adaptation of Greiner's model [36].

upheaval occurs in organisation practices'. The term revolution covers 'those period of substantial turmoil in organisation life'. The model has been modified by the addition of a 'crisis of sustainability' revolutionary stage in the fifth phase. Large, mature practices are often faced with a continuing battle to sustain their trading position on a number of fronts and to maintain strategic alliances in a changing market. They face periods of uncertainty which have been called 'crises of sustainability'.

Although much of this book concentrates on the development of a forward looking intuitive approach to marketing, design practices should still take into account their organizational history and current position in the growth cycle. Greiner notes that:

> Companies fail to see that many clues to their future success lie within their own organizations and their evolving states of development. Moreover, the inability of management to understand its organization development problems can result in a company becoming 'frozen' in its present stage of evolution or, ultimately, in failure regardless of market opportunities ... The critical task for management in each revolutionary phase is to find a new set of organization practices that will become the basis for managing the next period of evolutionary growth.

Design practices can become locked in one of the phases of growth despite the age of the practice. Members of the organization can be either unwilling or unable to move into the next stage especially if this involves a challenge to long held views and perceptions of directors or partners. Part of the strategic mapping process is to envisage the possible evolutionary and revolutionary stages en route to your preferred goal. Both evolutionary and revolutionary growth stages present many challenges no matter how large or small the practice and these stages of change need to be managed carefully. Greiner's model therefore offers one view of the likely stages on the way and shows how strategic mapping can be related to a preferred growth curve for your practice.

5.5 Strategic positioning

One of the classic market positioning strategies is to create a new category of product or service and thereby establish your organization as a market leader in a redefined market. In some ways this strategy has already been put into effect by other service providers in the construction industry. The redefined category of project management has been estab-lished by, among others, surveyors who have established market leadership in this market. Of course design professionals are still free to register their project management capabilities with prospective clients but in terms of market positioning they are now seen by clients as coming from behind rather than being the leader of the pack.

If design professionals are intending to reposition their services in a new market front, the selected market should meet a number of conditions:

- There must be a process of marketing entelechy which is clearly understood by both clients and service providers.
- It should be a market in which design professionals can offer a unique service and make a valued contribution.
- The market must be substantial enough to be worth the effort.
- Design professionals must be in a position to take the lead in the market and sustain their market leadership.

An example of a suitable market would be the refurbishment market. It is a substantial market which is likely to grow in the 1990s and beyond. It is also an area in which design professionals can identify a clear role and demonstrate specific competence. There is therefore an opportunity to establish distinct market positioning in an emerging market.

On the wider front, design professionals should be able to strengthen their position in a maturing cultural consumption market through a process of cultural positioning. As stated in Chapter 4, cultural positioning assumes that there is a cultural best fit between the company, the client and the sectoral infrastructure. Identifying and producing that best fit is the marketing task of cultural positioning. One of the ways of addressing wider markets could be through architecture centres which recognize and deal with the cultural consumption market. These centres are discussed in more detail in Chapter 7.

At the corporate level, companies are faced with the fundamental choice of what market they want to be in. To what extent should a practice diversify away from its existing markets and key strengths? Any decision to remain in a market or to diversify must be complemented by a review of marketing positioning. Restaging established services in new markets is unlikely to succeed. The reason is simple. It is wrong to assume that client perceptions of your company will be the same in different sectors. Restaging established services would be based on this wrong assumption. What a company must do is to ask a number of basic questions about the market positioning of its service:

- What is the market positioning message that will help to establish sufficient marketing valence with a new market?
- Does that positioning message accurately reflect the marketing composition that is being offered to either new or existing clients?
- What are the synthesis factors of current or planned services that would strengthen marketing valence?
- How will market positioning change throughout the process of marketing entelechy?

Market positioning is the pivotal marketing issue that will determine the success or failure of design professionals in the future. If architects and engineers do not take a more proactive and pre-emptive stance, they will be marginalized by the positioning activities of other service providers. If this happens, architects and engineers will continue to perform their basic design and technical functions, but will have less influence with clients and the sectoral infrastructure.

Any positive changes to market positioning at corporate and professional level must be gradual and controlled as part of the overall marketing agenda. However, progress cannot be made if market positioning is not even under consideration or is simply not understood by design professionals.

5.6 Vertical integration and disintegration

The cultural consumption market for design professionals is a fragmented market which is so large and diverse in its requirements that it needs a large number of practices to meet all the demands. In this type of fragmented market, vertical integration is a strategic option worthy of consideration. Vertical integration is the increase of control or influence in the chain of activity of the construction cycle by either takeover, merger, expansion or formal agreements with other companies. The intention is to integrate a number of stages in the process and thereby gain competitive advantage.

A vertical integration strategy in the construction industry will decide how much control and influence a company wants over the various stages of funding, design building and operation. This can be achieved by taking direct control of more stages of the process by takeover, merger or expansion or by agreeing in advance the market transactions that would occur normally between independent companies at different stages.

Each vertical integration strategy has to be looked at from the point of view of the organization applying the strategy. For example, the design and build procurement method is a vertical integration strategy from the point of view of the contractor. The benefit to contractors is that they have more contact with and influence on the client by extending their operation to an earlier stage within the construction cycle. Similarly, a contractor can form a strategic alliance with a funder or developer, set up or bring on board an operating company, and jointly provide a potential client with a complete package of funding, design, construction and operation. This would be full vertical integration.

5.6.1 The experience of vertical integration and disintegration in the manufacturing sector

The history of vertical integration and more recent vertical disintegration in the manufacturing sector provides some indication of the possible outcomes of vertical integration. Throughout the 1970s and 1980s large manufacturing organizations saw great benefit in vertical integration. For example, they felt that it would be advantageous to reduce their level of stocks of materials by having more control or influence over their suppliers. Suppliers were asked to provide parts only when they were required for assembly. The large manufacturers began either to set up their own supply chains or to invest in a number of separate dedicated suppliers. A typical example might be a large manufacturing company which had forty or fifty suppliers of a particular set of parts for assembly. That number would be reduced to four or five suppliers who would be expected to improve their efficiency and methods in return for a long-term, guaranteed supply of work. This was good news for the few companies which remained in the supply chain but bad news for the others left in the cold. At the other end of the cycle, the manufacturing company might want to take more control over the supply and distribution network in order to make sure that its goods reached the right market. Either way, there was a level of vertical integration which reduced the total number of companies in the process and increased the control and influence of the large manufacturing company.

More recently, large manufacturing companies have recognized that the operational problems of vertical integration, i.e. having to cover more stages of the processes, begun to outweigh the competitive advantages. There response has been to hive off parts of their organization. The newly autonomous companies have trading links back to the original parent company but have to trade in wider markets to continue to exist as viable organizations. This process has been called vertical disintegration. A major contributing factor to vertical disintegration has been the ability of organizations to maintain links using modern information technology. Information technology has become the vertical integrator that allows independence but supports interdependence.

5.6.2 Vertical integration in the construction industry

In the construction industry, the oversupply of both service providers and contractors has led to a increased competitiveness and volatility. At the same time, many larger client bodies which are also part of a supply chain, such as the privatized public utilities, are taking steps to integrate vertically. Service organizations, such as architects and engineers, are being asked to enter into for example, framework agreements in which they are given a preferred supplier status but have to agree to improve their performance over a set period. As part of the deal, public utilities offer training to the staff of the supplier. This is also one way of ensuring that the supply staff understand and comply with the organizational culture of public utility.

Another vertical integration option is 'partnering'. Partnering is defined by the United States Construction Industry Institute as: 'a long term commitment between two or more organizations for the purpose of achieving specific business objectives by maximising the effectiveness of each participant's resources'. Partnering is a term given to a formal relationship between a service provider, or service group, and a customer. For example, contractors can 'partner' with a large client body. The intention would be to reduce costs and improve quality levels over an agreed period. Similarly, a consultant could partner with other organizations to provide a joint service for a particular project (project partnering) or for a number of commissions (strategic partnering).

According to Bennett and Jayes [37], the full benefits of partnering require a number of conditions to be met:

- First there must be an acceptance by the partners that it is worth making the investment in building a partner arrangement. This requires commitment from top management and the delegation of appropriate authority throughout the whole organisation.
- Second, there must be potential for improvement in the product or service which is the subject of the partnering arrangement; and also the companies involved must have the potential to improve their performance.
- Third it must be recognised that for the full benefits of partnering to be achieved, it has to be medium to long term strategy because it

takes time for benefits to emerge. However, major projects that take several years to complete can use project partnering to achieve significant benefits.

The competitive advantage offered by partnering has to be set against the difficulties of team building and cooperation. However, in the remainder of the 1990s, partnering will continue to be attractive if it can be seen to deliver the claims of reduced risk, less competition and improved quality.

Individual architectural and engineering practices are also able to benefit from vertical integration. One way is to form strategic alliances with other service providers before or after your practice's design input in the construction cycle. Offering a different, related service either before or after the core design activity is another way. However, few practices have the corporate clout that would allow them to have direct control over the processes. What needs to be cultivated by design professionals is an increased level of influence which is spread as far as possible throughout the process from concept to completion.

A number of questions need to be asked about the long-term impact of vertical integration. Is widespread vertical integration in the best interests of design professionals? If vertical integration leads to a reduction in numbers of design practices, is the cultural diversity represented by these practices a positive aspect of a fragmented construction market that needs to be maintained? Will vertical integration lead to a reduction in the level of influence of design professionals in the design and construction cycle? Only time will tell if vertical integration will yield the competitive advantages envisaged by strategists of large organizations. In the interim, however, it could do more harm than good to the collective fortunes of design professionals.

5.7 Networking

A network is an arrangement, either formal or informal, by a number of architects and other complementary service providers, to market their services jointly. The purpose of operating a network is to widen the marketplace both geographically and in terms of type of new projects. The network exists to bring together complementary disciplines in order to benefit from synergy – the whole is greater than the sum of the parts. By doing this, practices can make joint bids for work that they would not have considered on their own. Networking is therefore a marketing strategy which extends the boundaries of operation of the respective companies involved in the network.

The practical operation of a network is based on:

- a commitment by each organization or individual to undertake developmental and speculative work on new services and projects. Each company needs to show that it can allocate resources to bid for and undertake the work;
- a clear understanding of which company in the matrix is taking the lead from the beginning of a potential project;
- a clear understanding of the purpose of the submission and why the work is being undertaken in strategic marketing terms.

As discussed in earlier chapters, the market reach of architectural services is considerable. If your company is therefore considering any form of networking, it may be worth looking further away rather than nearer to home. The final choice of market will depend upon the demonstration that there is a market need for your service in the chosen location rather than the distance from your existing practice.

5.8 Public sector procurement: legislative changes

In the Transart story in Chapter 3 there were several references to fictional legislative changes. These sort of changes have an impact on strategic mapping and in this section of the book there will be an examination of the impact of actual changes to public sector procurement legislation in the early 1990s. In this case the changes were to European directives that were proposed and agreed at European level and implemented within the national legislative frameworks of individual EU member states.

During the last few years there have been many changes to the structure of public sector institutions and organizations. These structural changes have taken the form of a devolution of powers, changes in incorporation and a corresponding increase in autonomy. Examples of these changes are the establishment of health trusts, creating competition in the public utilities and the change of status of former polytechnics, now universities.

Many changes have been undertaken under the heading of privatization. The general aims are to make individual organizations responsible for their budgets, to encourage private/ public sector partnership or to facilitate wholly private sector funding. The underlying rationale is that the dominance of private sector funding in the construction market of the 1980s will be replaced by private sector reliance in the 1990s. In the last few decades, most funding for the construction industry has come from the private sector. In the 1990s and beyond, there will be a reliance on private sector funding before projects, such as major infrastructure programmes can go ahead.

From the point of view of design professionals, the legislative changes to the process of public sector procurement have already had a significant effect. These changes have had an impact on the way in which architects and engineers are asked to bid for work, the range of services that have to be offered to win a commission and the level of financial remuneration.

Of particular importance to architects and engineers have been the implementation of EC Services Directive 92/50/EEC from 1 July 1993 and similar legislation for public utilities services contracts covered by Directive 90/531/EEC. Services Directive 92/50/EEC covers all public sector bodies except public utilities. (The public utilities are organizations in the water, energy and telecommunications sectors.) The definition of public sector in the legislation is all work 'financed for the most part by the State'. This definition is interpreted as all work that has more than 50% public sector funding and includes many, if not all of the commissions awarded by the public sector institutions and bodies that have been privatized throughout the 1980s and early 1990s.

Generally speaking, the Services Directive 92/50/EEC asks that all public sector architectural commissions, 'the estimated value of which, net of VAT, is not less than ECU 200,000', be advertised throughout EC member states in the supplement to the Official Journal of the European Communities. This supplement is a daily publication available throughout Europe. The process is similar to the Works Directive that has been in operation since the early 1970s.

Each public sector body has the option of undertaking the work in house without having to advertise. So the first step is to determine if the work has to be done by outside architects. (This situation is likely to change when further national legislation comes into effect and further opens up the public sector.) There are several different procedures that a public sector body can then adopt: namely open, negotiated and restricted procedures and design competitions.

The overall number of submissions that have been required of architects, engineers and other service providers since the implementation of the legislation has been very large indeed. For example, shortly after the introduction of the procedures, submissions for architectural commissions were totalling 150 to 200 for each advertisement in the United Kingdom. Although these numbers have dropped to between 50 and 70, and sometimes less, the sheer volume of work undertaken on submissions is considerable.

The most popular process of procurement is the restricted procedure. In the restricted procedure, the public sector organization can chose between five and twenty applicants for a shortlist from the total number of applicants. However, the criteria for selection are clearly defined in the legislation and only a few criteria can, in theory, be used as a filter in this first stage namely:

- basic criteria for rejection of suppliers such as bankruptcy;
- information as to economic and financial standing;
- the ability of the service providers 'to perform services with regard to their skills, efficiency experience and reliability';
- the production of certificates for attesting conformity to quality assurance standards.

Once the shortlist had been made, other criteria can be used but they must, in theory, be stated in the original advertisement. At this final stage, there is a level playing field and the final selection can be based on the 'most economically advantageous tender'. The 'most economically advantageous' criteria allow aspects other than simple cost to be taken into consideration. These criteria have to be stated in the first Official Journal advertisement in priority order by the client body.

Under Article 33 of the Services Directive, the only quality standards that can be referred to are BS5750 or the EN equivalents. The relevance of the BS5750 standard, which has been developed in the manufacturing sector, to the provision of architectural services, has been the subject of much debate. The idea that compliance with BS5750 will now become a key determinant in a public sector shortlisting process has been equally contentious.

Anecdotal evidence from decision makers in leading local authorities, suggests that the legislation has created more problems than it has solved. For example, many local authorities are still concerned about how to handle the process. In particular, they are concerned about how to reach a final shortlist from potentially large numbers of respondents to the EU advertisement given the limitations of the selection criteria and the time limits. The issue for many public sector organizations is that they are put to a great deal of trouble and inconvenience and in the end may not get what they want. The feeling in many organizations is that the final shortlist for a specific commission reached under this procurement system is unlikely to be the same as the shortlist produced by a public sector body free to use its own methods. The difference is dictated by the process and not by choice.

From a strategic mapping point of view, the changes to procedures present as many opportunities as threats. The obvious opportunity is for British design professionals to exploit market reach and win work in mainland Europe. The matching threat is that overseas architects will have increased opportunity to win work in this country. There are precedents for the latter situation. For example, Italian architects have already won work in major conurbations in this country. Their fees are reputed to be low and they have demonstrated a track record of delivering good work.

It is suggested that the less obvious opportunities for architects and engineers lie in the process of appointment. For example, there is a need to provide the public sector with advice in the pre-tender situation. This advice can be in the form of feasibility studies, property audits or even advice about the procurement process. Design professionals are well positioned to undertake this work.

Another opportunity lies in the considerable time required to make an appointment. Many public sector bodies are changing the way in which they procure services. The time taken from concept to appointment can, in theory, be as long a six to nine months. It is therefore not practical to appoint an architect and then begin to appoint other service providers. Potential public sector clients may be forced to adopt more comprehensive, quicker forms of procurement which place design services in a basket of other services. The result can be a change of role for design professionals. Architects and engineers are therefore presented with an opportunity to offer a wider range of services and to reassess their market positioning in order to maintain their place in the procurement chain. In many ways this is not a bad idea. Frank Duffy commented in his *Strategic Review of the Profession* [15] that 'The exclusivist habit must be reversed. A rich mix of old as well as new skills must be on offer to clients.' This sentiment is echoed in a 1992 report for the UIA called *Towards a New Architectural Practice*. The report was written by Weld Coxe and his wife Mary Hayden [38]. They conclude that by 'taking greater responsibility for all the myriad inputs that are required to carry out work', the architect's 'position in society is enhanced' and that this does 'not in any way diminish the quality of the work'.

The issue of procurement is a critical area that needs to be monitored continuously and addressed by professional bodies and individual practices if the outcome of procurement legislation is to favour design professionals. Strategic mapping is required at a number of levels to ensure that the balance of the equation of opportunities versus threats is favourable.

5.9 Alternative strategies

One of the threats posed to individual practices by changes to procurement methods is a possible long-term reduction in the amount of work won in any particular market. The reason for the reduction is that there is an increasing amount of competition in both public and private sectors. The changes have been enforced and accelerated by legislation in the public sector but more rigorous forms of procurement are also being adopted by private sector organizations. In the past, if you won a commission for a particular building type or structure, there was some guarantee that this would lead to other, similar work. However, if your practice is competing against a large number of practices for every subsequent commission, the same may not be true.

From a strategic mapping perspective, you have to take the possible reduction in the level of subsequent work into account. If you are potentially getting less work out of one sector, you have to maintain a steady stream of work from other sectors. There is therefore some pressure to consider diversification strategies which will bring in the required level of work in the long term. In practice, it could take a five- to eight-year period before you could come to conclusions about the success or failure of these diversification strategies based on outcomes. In the interim, the strategic map must take all of these issues into account. Retrospective analysis of market history will not be able to forecast likely outcomes and you should adopt a synthesis marketing approach if you want to formulate possible diversification strategies. You are faced with the challenge of not only maintaining a presence in a number of markets but also continuing to be considered a serious player in these markets by both potential clients and the respective sectoral infrastructures.

The increase in competition within each market means that fee reduction becomes an important differentiation issue if you are trying to maintain your place in an existing market. Unfortunately, however, nil or low fee bidding at the early stages of a commission has become a feature of the 1990s. What impact does such a strategy have on your practice's market positioning? What message does it send to the sectoral infrastructure and the outside world about the sort of professional services that you have on offer? In my opinion, nil fee bidding and low cost entry will erode your market positioning and will reflect badly on your professional status. Tight fee bidding and cost controls are a necessary part of winning work but the fee level should still reflect your position in the market.

Another strategic response might be to increase the geographical coverage of existing services. In order to make an informed decision, you have to balance the risk of reduced work in existing sectors with the costs and risks associated with the your chosen strategic response. Modern communications have reduced the risks associated with widening geographical coverage and if you are offering an existing service further afield, you have the comfort that you can point to a relevant track record of work. The synthesis marketing task in such a diversification strategy is to find ways of becoming accepted in the sectoral infrastructure further afield and to break down cultural barriers if you are moving into another country. If you are trying to break into a similar market in another country, it is not recommended that you market your practice as a provider of better core services. You have to market design concepts rather than design capability. The development and positioning of the augmented service needs more attention if you want to stand out from the crowd further afield.

A synthesis marketing programme

6

Perspectives are altered by the fact of being drawn. (John Updike [39])

Up to this point in the process we have:

- produced up to four scenario plans that provide a long-term set of options for the practice;
- formulated related strategic maps that help to determine the best way forward;
- stated a strategic preference.

Your next task is to translate these long-term strategic views into a synthesis marketing programme capable of delivering successful outcomes. There are numerous ways to translate potentiality in actuality (marketing entelechy), to deliver the requisite level of marketing composition and to ensure sufficient marketing valence. In this chapter a number of different marketing activities are examined which could form part of a synthesis marketing programme. A synthesis marketing programme is a series of measures that translate strategic aspirations and proposals into action. Six months is the recommended time period for the programme because it is usually the amount of time taken before events change significantly and is therefore the suggested time period before undertaking an in-depth review of the marketing strategy.

The duration of the synthesis marketing programme can vary between services and sectors but an agreement should be reached about its length at the beginning of the process. The review should be undertaken at a fixed date even if nothing appears to have changed. It is wrong to assume that everyone will have a common perception after several months' marketing and you need to make sure that the review is fixed firmly in the diaries of contributors to the marketing effort.

6.1 Marketing information systems

6.1.1 From client contact to winning the commission

In order to put a synthesis marketing programme into operation you need to have a marketing information system which:

- provides you with information about what is happening in selected markets;
- provides details of previous commissions of all types;
- collates information about potential new projects that are in the marketing funnel;
- monitors the client's experience during the commission;
- maintains contact with previous clients.

You can waste a great deal of time and energy collecting information about what is happening in the wider world. This information can come in many forms including press cuttings, reports, market research findings, and rumours about work that is likely to come up or which of your competitors is winning work. One of the benefits of having spent so much time on strategic issues is that you able to focus on what is relevant to your strategic goals. You should be in a position to maintain a clear focus about what is and what is not relevant to the future of your practice.

One of the most difficult skills to learn in marketing is the ability to keep a clear focus on marketing effort and not to expend vast amounts of energy chasing every opportunity. Sound tactical development is as much about focus as it is about methods. You have to learn to home in quickly on what is important and to let go of unsuitable projects. Unsuitability, in this context, is defined by the inability to meet the criteria established in your strategic map and scenario plan. The project that appears as an opportunity might be packed with personal interests and attractions but if it does not fit with your practice aspirations you have to let it go.

Once a project is felt to be of interest to your practice you should make a record of what it is about, who are the main contacts, and which people are in the decision making unit. You should also begin to file all the correspondence that takes place about the prospective job and introduce the project into the wider systems within your office. You can buy computer software that provides an integrated record from identifying the opportunity to completing the commission. If the system on offer suits the way you work then that is fine. However, in my view you can meet the requirements of even a large practice by a simple computer record or even a hard copy file card system for a smaller practice. What is often more difficult is getting accurate, detailed information about previous commissions that the practice has undertaken. You need this information to demonstrate your track record and to establish trust and confidence with the client in the early stages of contact.

Having access to information about previous commissions is especially important for large engineering practices whose work might span several continents. Engineering managers should have computer access to a database of previous work. The problems that arise in setting up the database are knowing what level of information is likely to be of interest to future clients and keeping the database up to date. In many cases potential clients are only interested in work undertaken in the last three or four years, and a vast database of previous commissions is not always appropriate.

In order to work out the scale of effort required to record and monitor potential commissions in the marketing funnel you need to do some simple arithmetic. Let us assume that your practice needs ten projects a year of all sizes to stay in business. You know from previous experience that if your practice is shortlisted for work you only win a commission on about one in six occasions. You also know that you are only shortlisted for about a third of all potential new jobs that you identify. This means that you have to identify, record and monitor about 180 projects throughout the year in order to deliver ten commissions. It is not suggested that these figures and rates are universal or even typical. The point to note is that there is a marketing iceberg in which the visible successful outcomes represent only a fraction of the hidden overall effort. If the marketing task of finding 180 opportunities a year seems to be too difficult, then the person responsible for marketing in this hypothetical practice will have to find ways of winning more commissions when short-listed or homing in on fewer possible projects in the early stages. Both courses of action present difficulties. At first sight it would seem logical that making better presentations at the final stages of shortlisting would be the best way of improving the ratings. However, paradoxically, it is often the case that reducing the number of projects under consideration brings a greater chance of success. The point is that by being more focused and selective you have probably begun to identify client requirement more clearly at an earlier stage. The result is that you are in a position to improve the contact and engagement with the potential client over a longer period.

If you feel that you cannot generate enough job leads to support your practice, and there are no reasons to suggest that you are merely missing opportunities, then you should go back to the strategic mapping stage and look again at the alternative strategies discussed in Chapter 5. If you feel that you have too large a rolling programme of possible job opportunities, you should see if you are casting too wide a net in too many sectors. At both ends of the spectrum the final result is the same – a failure to deliver a set of commissions that will meet your plans. Success in the marketing funnel is a result of balance and focus and you can only manage the process if you have established a marketing information system.

6.1.2 During the commission

It is often the case that the people responsible for marketing the practice have no further contact with the client once the commission has been won. The job of marketing is defined as the ability to win commissions and not the ability to engage with the client in an enduring relationship and to convey that experience to the sectoral infrastructure, i.e. the synthesis marketing approach. In synthesis marketing we want to extend the contact with the client without interfering with the day to day delivery of the service.

As a marketer you are trying to get alongside the project architect or engineer representing your practice and keeping yourself informed about the client experience. One way of doing this is to agree a promotional programme with the client, other funders and other service providers. The promotional programme is a listing of key events that are felt to be newsworthy or have special interest to outside groups. The list does not need to be extensive and should not involve a great deal of extra work. The intention is to find out what messages the various organizations involved in the project want to give to the outside world. There may be as few as four or five events that are worthy of mention. For example, events could include the start on site, the erection of the structure, and the period leading up to opening. Ideally you should arrange a meeting, chaired by the client, with all the organizations involved before the the work starts on site. The key events and their corresponding dates can be listed and you can get a better understanding of how the story can be told. What you have to avoid is taking on the role of promotional support for the client. If you feel that the client needs support then you can advise him or her accordingly. It is then the client's job to find the necessary resources.

If you are unable, for whatever reason, to maintain direct contact with the client, you should take steps to ensure that you have access to project planning information or feedback from the project architect or engineer. As a marketer you should examine the information about the progress of the commission from the client's perspective. Your intention is to look for any possible mismatch between the client's and the practice's perceptions, expectations and experience of the service. The mismatch is not always a sign of poor delivery. It is more often a sign of poor communication

6.1.3 After completion

Once the commission is finished, you should update your client records as personnel change within the client organization and you should keep a record of the commission in a form that is accessible and understandable for future marketing effort. If you invite clients to events or exhibitions try to make at least a mental note of ensuing contact. In my opinion practice newsletters sent out to former clients are a waste of time, money and effort. Clients are interested in the account of their project but have only a polite interest in the rest of your work. The best way to keep in touch with clients is often through the activities and networks of the sectoral infrastructure and not by direct contact.

6.2 The marketing role

The people responsible for marketing effort in design practices appear in a number of guises including:

■   design professionals who have acquired marketing skills and have taken on the role of marketing director or partner on a full- or part-time basis;

■   full-time marketing professionals who have come from a variety of backgrounds in other service or manufacturing sectors. They may be employed as marketing directors, marketing managers, business development managers or marketing coordinators and are responsible for providing a marketing support service for the practice. In some cases this support service could be limited to only marketing services with service and internal marketing not included in their responsibility.

The almost universal experience of people in these marketing positions is that their role is advisory and directional, not prescriptive. People in marketing roles, even at the most senior director or partner level, have had to learn to make suggestions rather than enforce plans. For some people this has been a difficult experience which is repeated in other professional services. Professional service organizations in general are predominantly flat structured, consensus based organizations that are felt by some writers on management theory to be a model of future company structures. If this is the case then many marketing people will have to put their powers of influence to the test in their own companies.

Another important quality that the person responsible for marketing must possess is the ability to pass ownership of the project over to the relevant design professionals. At the point of winning the commission, the architect or engineer who is actually going to undertake the work has to be sat in front of the potential client. The marketing person may have spent weeks or months setting up the situation but at the final hurdle they have to let go. Again some people find this a frustrating and unrewarding experience. The positive conclusion that can be reached is that if the marketing role within design practices is to be performed well, it will require the above skills and qualities to be part of the marketing discipline. In order to deliver the marketing effort in this way, the related skills of internal marketing have to be developed by those people responsible for marketing. In this mode of operation, internal marketing skills become as important as the skills involved in marketing services or service marketing.

6.3 Exhibitions

Not all sectors have exhibitions that are suitable for the promotion of architectural and engineering services. However, if they exist, they should be given a pre-eminent position in a synthesis marketing programme for a number of reasons:

■   They represent a way of dealing with both the sectoral infrastructure and existing and potential clients all in one event.
■   All forms of visual, audio and written of promotional media can be used to convey your message.

■   There is an opportunity to have direct face to face contact with potential clients in a neutral environment.

■   Many of the issues that face your chosen sector will be raised at either a linked conference or in the process of the exhibition.

■   They can show you what some of your competitors have on offer.

■   You can become aware of projects that are under consideration and there is often an opportunity for pre-emptive marketing.

■   A new service on offer can be either launched at an exhibition or can be presented to an invited audience.

In order to select and commit your practice to a particular exhibition, you should take a number of things into account:

■   Will members of decision making units be present at this exhibition?

■   Is the exhibition focused enough on your chosen market and its sectoral infrastructure?

■   You do not want to exhibit to delegates and visitors whose interests range too far and wide for your practice.

■   Are exhibition costs too big a drain on your overall marketing budget?

■   Which organizations are exhibiting around your stand in the exhibition hall? Remember that market positioning is as important as client contact.

■   If you are exhibiting at an exhibition for the first time, ask for a summary of the type of delegates and visitors that attended previous exhibitions. A full attendance list might also be available on request.

The build up to the exhibition is as important as the actual event. The marketing task in the period leading up to the exhibition is to remind as many people as possible that your practice will be present at the exhibition. This is a stage in the process that is often overlooked and it is easy to assume that you will meet everyone on the day. In fact the actual number of face to face contacts made by even the most active representative on exhibition stands is a relatively low compared with the potential number of people to be seen. You need therefore to spend as much time as possible inviting people to your stand and asking them to confirm their attendance before the exhibition takes place. When you arrive at the exhibition, you should have a detailed list of people who have been invited and corresponding list of confirmations. When a potential client, who has received an invitation, arrives at your stand, the person making the first contact from your practice should show that his or her name, interests and background are known and remembered. This means that everyone on the stand has to be well briefed. Preparation for exhibitions is as important as delivery on the day.

The mistake that is often made by practices attending exhibitions is to define the success or failure of the exhibition only in terms of job leads or level of contact with potential new clients. These aspects are of course important; but the real measure of success of exhibitions is their ability to establish or reinforce market positioning within the sectoral infrastructure. All of the advantages listed earlier make exhibitions an ideal promotional medium to make a market positioning statement as well to win work.

Another mistaken view of exhibitions is that they are an expensive form of marketing. The superficial view is taken that a relatively large sum of money is invested in an event that can last only one or two days. Hotel and travel expenses are incurred, stand equipment can be expensive, and time, money and human effort needs to committed to produce a good exhibition stand. However, you have to compare these costs with the costs involved in making contact with the same audiences and individuals over a longer period of time. Exhibitions are almost always a cheaper option if they were looked at in this way. Indeed, you may come to the conclusion that you could not easily make contact with these individuals and groups by any other method than an exhibition.

6.4 Cold calling

Cold calling is the common expression given to initiating contact with a potential client for the first time. It is a misleading expression because although contact is starting from cold, the contact experience is not necessarily frigid. Similarly, 'calling' implies a visit to the client or a telephone call. Again neither interpretation is accurate because you should try, where possible, to precede any form of contact with a letter explaining your reasons for having a meeting or telephoning.

If you are to attempt any proactive or pre-emptive marketing then you will at some point need to cold call potential clients. Indeed, an element of cold calling should be built into your synthesis marketing programme. However, cold calling should not be done without a reason. You might, for example, have heard that the prospect has been awarded funding for a project. You might then decide to make a telephone enquiry to the potential client in order to confirm the rumour and to express an interest on behalf of your practice in undertaking the work. If you come from a position where you are offering something of potential benefit to the prospective client and you are interested in what is happening in their world, then cold calling becomes an easy exercise. Alternatively if you come from a position where cold calling is beneath your professional dignity and you are somehow chasing an uninterested, resistant client, then an unhappy, unsatisfactory experience is likely to result. The shared view of many marketing professionals is that people are willing to talk about their projects and are interested in what is on offer. Much of the success of cold calling depends on your attitude as much as the potential client's. The only major pitfall to avoid in cold calling on the telephone is that you should not try to make detailed points about your practice capability, unless requested. Details should be covered in either a letter or a later face to face meeting. In a face to face meeting you are able to judge reactions in a way that is almost impossible on the telephone, even for an experienced marketer.

One of the most interesting experiences in marketing is the first meeting with a potential client. No matter how well prepared you are for the meeting or how well briefed, you can never predict what will happen. Finding out what is happening in the world of a potential client at a face to face meeting is a fascinating experience. Engaging with the client and hearing first hand about their plans and aspirations, albeit in an edited form, can be both illuminating and exciting. The development of a positive attitude to cold calling can quickly lead to a better understanding of a market outside your contact with existing clients and can be lead to increased opportunities of winning work.

6.5 Handling the media

In the past, many design practices measured their marketing effort by the amount of press coverage they received. Marketing effort was limited to marketing services and the measure of performance of marketing services was often taken to be the number of recent press releases and resulting press coverage. This is a very narrow view of the potential opportunities that exist in both the traditional media of press, television and radio as well as the emerging opportunities that exist on the Internet (discussed later in this chapter).

6.5.1 Radio

The most undervalued medium for the promotion of design services is local and national radio coverage. In the cultural consumption market, there is a need to stimulate and excite the imagination of both clients and end users. Radio coverage is a way of inviting these groups to use their imagination and to share the thoughts of the designer in the intimate environment of the radio studio. The imagination of the radio listeners is not limited by the presentation of visual images but is being extended without boundaries by the experience and thoughts of the design professional. If the ideas of the designer are presented in straightforward manner with no technical language, then the result can be an excellent communication of ideas and a better presentation of the project from the designer's point of view. If you are putting together a promotional programme for a project, try including radio coverage. Radio stations welcome special features and you might be pleasantly surprised with the end result.

6.5.2 Television

The opportunities for design professionals in the medium of television come in three forms:

■   involvement in a full-length television programme that is about design or related issues;
■   contribution to a television programme or news feature that has a slot for a story about a particular building or structure;
■   an interview on television as part of a news item or specialist programme.

Only a handful of architects and engineers are involved in the production of full-length television programmes each year. However, there are an increasing number of opportunities for design professionals to contribute to special features or to be interviewed about a particular subject. The main difference between the two types of television appearance is the time on screen. A special feature might last two or three minutes and an interview is unlikely to last more than a minute. Both opportunities have potentially high visibility and impact upon a public audience and demand a similar sort of discipline if the feature or interview is to be successful.

Like it or not, we live in a sound bite era. The ability to argue a case in two or three sentences is a discipline to be developed if you want to address a television audience. In an interview situation your contribution could be limited to only a few sentences. If you appear in a feature item you might be able to have, say, a short question and answer session with a television interviewer. Your replies would again be in two- or three-sentence blocks.

The person from the practice who will be appearing on television is faced with two tasks. He or she has to marshal the arguments in small, non-technical, digestible chunks and become familiar with the workings of television, in particular the operation of a television crew. When a television interview or feature is broadcast, all that appears on the television screen is an image of the person being interviewed and perhaps a view of the interviewer. In reality, the person being interviewed is looking at the faces of at least five or six members of the television crew. The point to remember is that the person being interviewed is in charge of the interview, not the interviewer. If, for whatever reason, you want to stop the interview then simply say 'cut' and stop talking. Believe me, it works. Always remember that you are in charge of the interview. Even professional presenters forget their lines or make mistakes and camera crews are used to retakes.

The medium of television therefore presents a high impact marketing opportunity that requires tight preparation if your message is to come across effectively. The opportunity to present your case exists in a relatively short space of time and detailed preparation is needed if you want your message to come across in such a short time period. The final outcome can be well worth the effort.

6.5.3 The press

In the world of marketing 'self praise is no recommendation'. Press relations avoid this trap as they represent, at least in theory, a third party view of the performance or methods of a practice. Press relations are therefore an important ingredient in the synthesis marketing effort.

When dealing with the press you should never lose sight of the fact that your ultimate marketing aim is to address existing and potential clients as well as their respective sectoral infrastructures. Indeed, in many ways the press are a vital part of the sectoral infrastructure. Not only do they provide a medium for the presentation of your practice but they also are part of the network of contacts and referrals that lubricate the process of marketing entelechy.

Good press relations need to be cultivated over a long period of time and you can to establish good relationships by taking a few basic steps:

■ Ensure that you have chosen the right publications by asking for information about the circulation, distribution and frequency of each publication and checking that it is being read by your members of your selected market.

■ Take time to speak to journalists or editors about what issues they feel are important to their readers. The suggestions for press releases should:
  – be simple and short;
  – be written in short jargon free sentences;
  – not assume that the journalist or editor has any specialist knowledge, even if it is a trade publication;
  – stick to the facts and avoid superlatives;
  – highlight the areas of possible interest to the readers of the publication;
  – summarize the story in the first few sentences of the release.

■ Feature articles written by members of your practice should follow all the guidelines for issuing press releases but should also:
  – be written as an exclusive article for a specific publication. You are giving the publication and their readers a unique insight into your experience and views;
  – avoid any reference to the business activity or commissions of the practice unless it is agreed by all parties that they should be discussed;
  – lead the reader through the story. You want to keep the interest of the reader from the beginning to the end of the article.

■ In helping editors or journalists to write their own articles about your work or related issues:
  – you should try to get the journalist to give you a clear picture of what the story is likely to say. You have to make sure that there isn't a negative slant in the story or that your contribution will be misinterpreted or presented incorrectly;
  – the journalist should visit your practice if possible. This give you an opportunity to meet face to face and will help considerably your mutual understanding and in particular the journalist's understanding of who you are and how you operate;
  – you should not give one- or two-sentence quotes. Short quotes do not give you the opportunity to express your point of view accurately and they are a sitting target for misinterpretation.

Whatever the method of engagement with the press, press relations are an integral part of the synthesis marketing programme. The press are important both in terms of conveying a message to the outside world and as players in the respective sectoral infrastructures. However, press relations are not the primary driving force of marketing entelechy nor do they dictate the strength of marketing valence. Press relations play an important part in the marketing effort but it is wrong to assume that they are the sole measure of performance of marketing effort.

If you feel that you need support to deal effectively with the press, then you might consider employing a public relations consultant. The promotion of design services requires specialist PR support and you should shop around for PR consultants that at least have an understanding of the construction industry, and better still direct experience of working with architects and engineers. Bear in mind that the duration of engagement with a PR consultancy is likely to exceed the review period for the synthesis marketing programme. PR support is not a quick fix solution that guarantees short-term coverage in selected publications. Indeed, the list of publications identified by a competent PR agency could well be different from you own preferences. For example, your favourite architectural and engineering publications might not feature high on the list of publications prioritized by the PR consultant. You are paying a PR agency to have a more objective view of what press coverage you need to achieve your aims. If your aims are not clearly stated in your marketing strategy or you are still in the process of coming to an agreement about a scenario plan or a preferred strategic map, then do not even attempt to appoint a PR consultancy. Most PR consultants are not in the business of sorting out your strategic issues, despite their claims to the contrary. They will find it hard to deliver their services if you cannot provide them with a succinct account of what you want them to do in strategic terms.

6.6 Brochures

Practice brochures are the reflection of the strategic plans that have been drawn up earlier. For this reason, brochures are not simply about pretty pictures, good design and fine words. They are a strategic statement about the practice. One of the problems in reaching an agreement about the style and content of brochures among design professionals is that there is a tendency to articulate the production of the brochure in only design terms. In the ensuing discussion about the form and content of the brochure, the real issues of strategic presentation and market positioning are often lost in the debate about colour, shape and size. It is suggested that if you want to produce a new brochure, try to list the strategic issues in short bullet points before you look at its design. It should be easier to produce the brochure if the strategic points are clear and not entangled in the general discussion.

If you produce a general practice brochure or statement, it should be used only for the first contact with potential clients and members of the sectoral infrastructure. If, however, you are seeking work in several sectors, you must take care that your general brochure does not in any way contradict or undermine your market positioning in the selected markets. If there is any doubt about client perceptions of the brochure, then it is better to have either no introductory brochure at all or to produce a separate set of brochures that are targeted at individual sectors.

Ideally, the early information given to a potential client should be in a form that will help to inspire trust and confidence and will if possible address the special needs of clients in their particular markets. If your general practice brochure does not help to achieve these aims, then there is a strong marketing argument that written and visual material should be held back until after the first face to face contact with the client. At this stage the subsequent submission can be tailored to suit the client. There is no golden rule in marketing that states that you must have a general practice brochure.

6.7 Conducting a client and sectoral infrastructure audit

If you want to establish a synthesis marketing programme targeted at existing and potential clients and their sectoral infrastructures, there is an implication that their perceptions and experience of your practice will be sought and recorded. This can be achieved by conducting an audit of client satisfaction and practice reputation in the sectoral infrastructure.

Conducting external audits is the one area of the synthesis marketing programme that should always be done by an outside agency, i.e. not in-house. The reason is that objectivity is important if the results are to have significance and usefully inform the marketing strategy. Frank and free discussion is needed to get any meaningful response from previous clients or members of the sectoral infrastructure who have had had any direct involvement with your practice. If the findings are to have any strategic significance, the audit should be undertaken by professionals who have particular skills in qualitative research and interview techniques. Ideally they will have conducted similar audits in the same or related markets.

The minimum requirement from the audit or audits is a better understanding of the synthesis factors that are important to both clients and members of the sectoral infrastructure. In addition you should be looking for qualitative information about perceptions of your practice and the issues and values of the market sector that will shape its future. If your practice has operated in a particular sector for many years there is an instant assumption that you will have an intimate understanding and appreciation of what is important in that particular market. However, in many cases, that perception is limited to a view from the practice's position in the sectoral infrastructure. It often takes an outsider to represent other points of view and to identify the important factors.

6.8 Direct mail (mailshots)

Direct mail is the production and distribution of letters to potential or past clients. The letters give details of your practice capability or services on offer. There is, however, a world of difference between the sort of junk mail that arrives on all our desks every day and a well crafted letter offering professional services addressed to the individual decision maker. The difference is often seen in the response rate to the mailshot. Scatter-gun junk mail has single-figure percentage response rates. Carefully worded mailshots that are targeted at individuals can have a response rate of between 20% and 50% depending on the market sector and the number of people being targeted. If you want to increase the response rate above this level you would have to either narrow the sectoral defini-tion and thereby reduce the number of people being targeted, or, in certain circumstances, offer inducements to reply.

There are two aspects of mailshots that are commonly not under-stood by architects and engineers. The first is an appreciation of the skill and methods required to produce a good mailshot. And the second is an understanding of what the practice wants to achieve by sending out a number of letters, possible with enclosures, to the selected group. In order to write a good mailshot you have to put yourself firmly in the shoes of the people receiving the mailshot. You can guarantee that the letter will be filed in the bin if you:

- make general statements about your practice capability;
- use technical terms with a layman;
- fail to state what benefit is on offer as well as the details of the service;
- do not make some reference to the key issues, problems or oppor-tunities that face the person reading the letter.

The advantages of mailshots are that they are a cheap, flexible way of addressing niche markets. At the lowest level of achievement, they can raise awareness of your services among potential clients. At best they can show you who is active in the market and who has projects in the pipe-line. If you receive a reply from potential clients stating, for example, that they have just appointed an architect or engineer to do similar work and they have no plans to employ design professionals in the foreseeable future, you can at least cross them off your target list. You have gained useful market information that will allow you to manage your time more effectively with a minimum expenditure of effort. The flexibility, cheapness and focus of direct mail means that it should be treated as a serious marketing option in any synthesis marketing programme.

6.9 Advertising

The strength of advertising is its ability to reinforce and support an existing market position. Its weaknesses are its high cost in relation to other marketing media and methods and its inability to turn market awareness into market acceptance. Being aware that a service exists in a particular market is not the same as market acceptance of your practice as a serious player and contender for future commissions. If the objective of your marketing effort is simply to raise awareness then advertising might be a consideration. However, my view is that advertising is less and less potent to address niche markets of interest to design professionals. If advertising is used it has to be for specific purposes on specific occasions. Its high cost has to be weighed against likely impact.

6.10 Surfing the Internet

Current estimates suggest that there are over 5 million host computers and 40 million people are now linked up to the global network called the Internet. The origins of the Internet lie in the US government Advanced Research Projects Agency network (ARPAnet) which began operating in 1969, using the new technology of packet switching. Packet switching breaks data up into small packets for travel across the network allowing multiple users to access networks at any one time. The packet software seeks to establish the fastest route to send each packet along the telecommunications network. The result is that if a section of network were rendered unusable (say from a nuclear strike) the data would still find a way to reach its destination.

In the early stages, ARPAnet grew at the rate of one new computer every month for about a decade. At the same time other networks were being established such as satellite and radio packet networks which ARPA wanted to connect to ARPAnet. At this time a new network protocol was formulated which could support connection across different networks known   as   Transmission   Control   Protocol/Internetworking   Protocol

(TCP/IP). In 1983 ARPAnet switched over to TCP/IP, an important move which many people date as the start of the Internet.

Around this time other networks were beginning to emerge which were independent of ARPAnet and in 1987 the National Science Foundation set out to establish a high speed backbone network to cope with increased network load. NSFnet was set up as a series of regional networks linking to the NSFnet backbone. NSFnet's original charter stated that it should only carry traffic for 'educational and research purposes', but at the turn of the 1980s new commercial organizations were springing up offering a connection to the Internet for anyone who would pay for it. These commercial providers offered a combination service and regional and backbone networking. However, users soon found that regional NSFnet links were refusing certain types of traffic.

In March 1991 the commercial organizations set up an independent high speed link between their own networks called the Commercial Internet Exchange (CIX). This single development revolutionized the Internet, allowing commercial access for non-military, non-educational purposes. During early 1993 the revolution was completed with the arrival of the World Wide Web. It was developed at the CERN Physics Laboratory as a way of sending multimedia documents across the Internet. It also had the added feature of hypertext linking between documents. Using hypertext links, users could click on a link, often shown as underlined text and the software would take them to another document, either on the same server, or on one at the other side of the world.

The early software for accessing the World Wide Web (WWW), known as browsers, was text based, but in November 1993 the first graphical browser appeared. NCSA Mosaic showed text, hypertext links and images all within the one window on screen, giving the appearance of an on-line brochure or magazine. With the highly successful launch of the Netscape browser in October 1994 graphical browsers quickly became the norm. This gave designers the possibility of producing brochure standard documents. These considerable improvements in technology and software has meant that the WWW has become the fastest growing part of the Internet. In mid-1995 Lycos, a search tool on the Internet, contained 3.25 million URLs (electronic addresses of WWW documents), and estimates suggest as many as 10 000 new URLs are added every week.

The Internet is thus a marketing and communications medium that cannot be ignored by architects and engineers. The Internet and the emerging information superhighway will revolutionize the process of marketing entelechy. Potentiality will become actuality in many novel ways, some as yet unforeseen, as a result of this communications revolution. The capabilities that the Internet offers include:

- Global mailing using the e-mail system. The same messages can be sent to a number of different locations on the Internet and can be stored until the person receiving the message asks for their mail messages.
- The multimedia (sound, text, video and graphics) capability of the World Wide Web allows design professionals to communicate their ideas globally.
- Any practice can set up a 'site' on the World Wide Web. The 'site' can use all the multimedia capability to convey any message about the practice or thoughts on any subject. The 'site' can be 'signposted' in other locations on the Internet.
- There are estimated to be over 20 000 newsgroups on the Internet. These newsgroups are for people who have a common interest and wish to exchange ideas. Newsgroups for architects and engineers exist already on the Internet.
- A telnet service exists to allow remote access to the computer that you have linked to the Internet.
- If so inclined, you can have real-time discussions with other people using the Internet relay chat (IRC) capability. There is also some indication that a videoconferencing capability will be available on the Internet in the near future.
- You can download information and software from the Internet using a file transfer protocol.

All of this existing and future capability of the Internet makes it a powerful low cost marketing tool that exploits the market reach of design services through every possible medium. The Internet should feature as a possible marketing option in the synthesis marketing programme of any practice, no matter how large or small.

Architecture centres: a marketing case study

7

The failures of contemporary planning and architecture are not the failures of architects or planners alone. They are also the failures of society and, in particular, of the resource controllers, the public and private developers who commission and brief architects, and of the system of controls, constraints and financial rewards or penalties within which architecture must be practised. If better results are to be achieved, then people must understand the context within which architecture is practised and where the real power lies to make or influence decisions. (Malcolm MacEwen, *RIBA Journal* [40])

7.1 The excluded middle

In April 1993, Rory Coonan, Peter Davey, and Rob Cowan produced a summary of the activities of ten architecture centres for *Architectural Review* magazine [41]. The architecture centres discussed in the article were:

- Pavilion de l'Arsenal, Paris;
- RIBA, London;
- the Zuiderkerk, Amsterdam;
- Arcam, Amsterdam;
- the Dutch Architectural Institute (NAI) Rotterdam;
- the Archicentre, Rotterdam;
- the Finnish Architectural Museum, Helsinki;

- the Danish Architectural Centre, Gammel Dok;
- the Canadian Centre for Architecture, Montreal;
- the German Architecture Museum, Frankfurt.

This short summary did not suggest that there was a 'simple formula for involving the public with architecture'. It sought, however, to present some of the best national centres as a basis for discussion and development and as a prelude to the Arts Council of Great Britain's conference on architecture centres Designing Cities which was held on 29 April 1994. The aim of the conference was to find new models of involving the general public with architecture and planning.

One of the outcomes of the conference was that the Arts Council of Great Britain agreed to part fund feasibility studies of architecture centres in a number of locations. By the beginning of 1994, nine studies had been prepared by the following organizations:

- the Birmingham Design Initiative;
- the Bristol Centre for the Advancement of Architecture;
- the City of Liverpool;
- the Architecture Centre, London;
- Hammersmith Community Trust, West London;
- RIBA Northern Region, Newcastle-upon-Tyne;
- the University of Nottingham;
- Plymouth Architectural Trust;
- Chatham and Medway Architecture Centre.

Later in the same year, the Arts Council commissioned Adrian Ellis [42] to review the feasibility studies and to make some collective observations. In his report, Ellis refers to what he calls 'the excluded middle' and suggests that:

> There is a tendency in the studies to move away from the high ground (What is an architecture centre? Whither Birmingham? etc) to the nuts and bolts of how much the Centre will cost and where the money might come from, and where to put it, without dwelling at too much length on the bit in the middle – the programme of activities itself and how to market it. There is a catalogue of functional components – exhibitions, databases, meeting rooms, cafes and bars – and some illustrations of the sort of thing that might be possible. But no real heart. This is important. As anyone running an arts, design or architecture centre will concede, the quality of the programme and how it is

marketed is ultimately more important than either the philosophy or the arithmetic.

In this final chapter of the book, there is a closer inspection of 'the excluded middle' and in particular a detailed examination of the marketing strategy for the proposed Northern Architecture Centre in Newcastle-upon-Tyne. In the latter case, the work on the marketing strategy is ongoing at the time of publication. However, the scenario planning and synthesis marketing effort undertaken to date has revealed a radically different view of the architecture centre concept, its mode of operation and its presentation to a number of audiences.

7.2 The Northern Architecture Centre: background

The idea of a Northern Architecture Centre was first put to a series of public meetings in 1993. The result was wide spread support for the concept. A steering group was established which in turn appointed a task force to drive the project forward. Over £50 000 was raised to establish the viability of the project and to appoint a project development manager. The task force and the project development manager were charged with the tasks of establishing the basis of the company limited by guarantee having charitable status which would manage the centre; negotiating the site procurement; gaining financial support from a number of sources; producing a marketing strategy; developing a business plan for the running of the centre; and holding an architectural competition.

When the first meeting of the Northern Architecture Centre steering group was held in October 1993, a number of questions hung over the assembled group namely:

■   What is an architecture centre?
■   Where can it be built?
■   Why is there a need for an architecture centre at this time?

Why, at this point in history, was there a perceived need to set up at least nine architecture centres in England and two in Scotland? Many of the reasons for the establishment of architecture centres have been covered in the preceding chapters of the book. Three reasons for developing architecture centres are singled out:

■   There has been a rush to improve the logistics of service and construction as a result of greater vertical integration, the productization of services, and the intensity of competition in an oversupplied market. It is felt that in this process the parallel concerns of design, purpose and value have been ignored. This is particularly true in the area of architectural service provision.

■ Design professionals are looking to find ways of dealing with a cultural consumption market by the establishment of a public forum for debate.

■ The social and economic restructuring that has occurred in the post-Fordist period has led to an undermining of the professional status of design professionals and consequently the design service that they offer. Architecture centres are seen as one way of redressing the balance.

In the case of the Northern Architecture Centre, a local architect, Geoffrey Purves, had been a long-standing supporter and champion for an architecture centre in the region and had drawn together four organizations with varying interests in the fields of architecture, community development, education, training and the construction industry. The intention was to build an architecture centre that would cover the northern region of Cumbria, Northumberland, Tyne and Wear, Durham and Teesside.

One of the potential partner organizations was the the Northern Architectural Association. The Association had existed since 1858 and its formation preceded the RIBA's. John Dobson had been the first chairman of the Northern Architectural Association and his proud tradition had been carried on by the Trust and learned society.

Newcastle Architecture Workshop had an international reputation as an architectural resource centre for the community. Its aim was to address the real needs of individuals and communities and empower them to play an active role developing their own futures.

The Centre for the Built Environment was a tripartite organization that brought together the considerable resources of Newcastle University, the University of Northumbria and Newcastle College. The Centre sought to offer an integrated service of education, training and career advice opportunities throughout the region and was one of the first of its type in the country.

The fourth potential member of the group was the RIBA regional office. It was felt that under the umbrella of the Northern Architecture Centre, the RIBA regional office could potentially operate in a different way. However, from the start it was agreed that no single professional interest should dominate the marketing agenda or determine the programme of future activity of the Northern Architecture Centre. Architects in the northern region would have to cooperate and collaborate with other professions in the construction industry to develop a non-protective, non-institutional approach supportive of all the sector.

7.3 The Northern Architecture Centre: feasibility study

At the end of 1993, the Steering Group of the Northern Architecture Centre commissioned European Economic Development Services Ltd to produce a feasibility study [43]. The main aim of the study was to look at the best practice of national models outside the UK and to suggest a model that would be acceptable to all the protagonists. There were no regional architecture centres in this country and a regional centre in the north of England would be a ground breaking concept. The study sought to be advanced but not prescriptive, to involve the various groups in a viable future and to refer to a possible wider agenda.

The feasibility study suggested a mission statement for the Northern Architecture Centre; namely 'To create the mechanisms in the Northern Region through which people can increase their knowledge, understanding and appreciation of historical and contemporary architecture, and in doing so raise their awareness and involvement in developing rural and urban environments for the future.'

The study referred to the accepted view of an architecture centre which was set out by Malcolm MacEwen in a May 1973 RIBA Journal article [40]. The feasibility study reports that MacEwen's idea of an architecture centre was based on 'the American experience of successful interpretation of landscape and natural environment in their national parks'. MacEwen argued that similar techniques of interpretation 'can be applied with equal success to an urban situation to architecture and the built environment'. He envisaged that a consequence of their establishment in the UK would be to enable the individual to participate more effectively in policy or decision making and to raise the standards of town planning. Quoting Freeman Tilden (the author of a standard text on interpretation in national parks, Interpreting our Heritage) MacEwen drew a distinction between information and interpretation: 'Interpretation is revelation based on information … The chief aim of interpretation is not instruction, but provocation.'

MacEwen concluded that the American experience in 1973 was one that had become accepted as relevant for Architecture Centres in Europe. MacEwen felt that:

The aim of architectural interpretation centres is to provoke curiosity, exploration and action, to arouse a sense of involvement and responsi-

bility. Interpretation should be concerned with the future as well as with the past, and must treat towns as living things, with problems both of conservation and change. So architectural interpretation centres are fundamentally different from traditional architectural exhibitions, which are ends in themselves: they are foci around which other activities can be developed. A visit to an architectural interpretation centre is the beginning not the end, of experience ... The function of the AIC is to 'tell a story' that explains what is to be seen outside, neither didactically nor comprehensively, but in the manner best calculated to provoke the desire to learn and to see more.

The feasibility study concluded that 'the MacEwen paper of 1973 provides a relevant context for the consideration of the role and function of an Architecture Centre in the Northern Region today'. However, after the later scenario planning workshop, it was agreed by the steering group that the activities of the Northern Architecture Centre would have to go beyond the basic provision of information, interpretation and provocation. The Northern Architecture Centre would have to respond to the funda- mental cultural, economic and social changes that had taken place in the last twenty years and in particular the changing role of design professionals in the construction industry. The interpretation and provocation model of 1973 was felt to be an inadequate response to the requirements of 1993. What would be required in the late 1990s would be marketing effort that focused on what MacEwen refers to at the beginning of the chapter as 'the context within which architecture is practised and where the real power lies to make or influence decisions'. The architecture centre of the future would have to deal with the primary concerns of execution and influence as well as the secondary issues of information, interpretation and provocation.

7.4 Marketing composition and repositioning

In terms of marketing composition, some of the potential ingredients for the Northern Architecture Centre were already in place in the form of the existing, potential partners in the project. However, each organization addressed its own audiences and clients in its own way. One of the early marketing tasks would be to examine the marketing position and possible co-positioning of the four partner organizations.

It was possible to envisage a scenario in which each organization was able to reappraise its existing activities in line with the aims and objectives of the architecture centre. This might not involve any radical redirection or reorganization. The most likely outcome was that the existing activity of each organization would be strengthened and expanded to match the task in hand. For example, there would be a need to provide a participatory and non-participatory programme of activity for the architecture centre. The Northern Architectural Association and the Newcastle Architecture Workshop might want to play a part in this programme. Similarly, there would be a need to promote the benefits of good design, high quality architecture and sound construction practice to the institutions and organizations that made up the local and regional economy. The RIBA regional office and the Centre for the Built Environment could take on this work.

There were two fundamentally different ways of dealing with the marketing composition. The first way was to put all of the activities of the selected parties together, see what they looked like as a composite body and then develop a statement of collective objectives that reflected their combined activities. The second way was to develop an outline statement of collective objectives, examine the existing activities of the various parties that would occupy space in the new centre, redefine and reposition the activities of these various organizations as far as possible to meet the collective statement of objectives, and then to note any gaps between the collective provision and the collective objective.

The first way would have been easier and quicker. The second approach, which was the chosen method, required more time and effort because it involved a more complicated iterative process. Representatives of the steering group would need to keep going back to each party with revised positioning statements and each potential partner would have to work on its own marketing plans and aspirations. Similarly, the second approach demanded a degree of consensus about an as yet unspecified collective objective. The advantages of the second approach were that it would provide a firm understanding by all parties of the nature of the synergy of their joint operation i.e. an understanding of what they could achieve together that they could not as individual organizations. It was agreed by the Northern Architecture Centre Steering Group that the way to develop statement of 'collective objectives' would be to develop a scenario plan i.e. a vision of what the centre would look like in ten years time.

7.5 Scenario planning: first stage

Throughout the process of development of the project, the steering group and the board of directors, established later, have used scenario planning methods to produce contingency plans for many eventualities as well as to maintain the focus required to deliver the project.

It was possible to take a view that the organization and marketing of the centre was fairly straightforward. The only requirements were that:

- the four organizations acting as the key players moved in having agreed basic financial and operational ground rules;
- a centre management structure be agreed;
- promotional literature be produced which reflected the activities of all parties.

So what was the problem? As mentioned earlier, the problem lay in the increased expectations of architecture centres. The need for a more sophisticated, wider approach came out of the first half day scenario planning workshop in May 1994. The workshop was led by a consultant, William Roe of William Roe Associates. About thirty senior representatives of key organizations and bodies in the region were invited to attend. Thirty was perhaps too large a number of people to make any inroads into detailed scenario planning but it did give an opportunity for people to describe their vision of the future of the Architecture Centre. By the end of the afternoon the level of energy and commitment to the centre had risen considerably and the aim of establishing a common view and starting point had been achieved.

After an initial discussion, the representatives attending the scenario planning workshop were asked to:

- describe their view of the centre in ten years' time;
- describe the driving forces that would shape and form the Architecture Centre project as it moved forward;
- list the positive and negative aspects of Centre's operation and impact;
- outline the method of alliance and ownership that would be suitable for the Centre;
- describe the activities of an up and running Centre;
- suggest possible names for the Centre;
- list the critical success factors of the project.

The long-term view of the Northern Architecture Centre that came out of the first half day workshop was an architecture centre which would:

- have an international dimension;

- develop a strong alignment between art, architecture and the environment;
- be a centre of excellence;
- be a focus for open, critical debate;
- advocate quality architecture;
- have the full support of the construction industry;
- have close links with local communities;
- influence building procurement in the region.

A tall order indeed. In a way, the Northern Architecture Centre was lucky. If the four participating organizations had not existed, they would have had to be invented to meet such widespread objectives. All four organizations represented distinct service and marketing agendas: no single organization would have been able to deliver all the aims and expectations of an architecture centre as envisaged in the scenario plan. The only drawback to having four existing bodies was that ownership of and commitment to the new project became pivotal issues.

7.6 The early marketing approach

The early marketing approach had to perform a number of functions. It had to be simple enough to introduce the basics of marketing in an easy way to a steering group whose members had little experience of or training in marketing of any kind. And it had to cover the two broad areas of consumer and business to business marketing. For example, the visitors to a new architecture centre represented a large consumer group that would have to be addressed separately from contact with community groups, educational institutions and local businesses.

Although the traditional marketing methods for consumer and business to business markets are different, a common method of approach was adopted for convenience. The approach is illustrated in Fig. 7.1.
It includes:

- a determination of broad market areas and their backdrop;
- detailed market research and analysis of each market;
- a breakdown of market segmentation. Market segments are homogeneous groupings that reflect the activities, preferences, and profiling of the members of each group. For example, in the tourism market, this would include the profiling of separate visitor groups and in the procurement market it could be a list of suppliers of certain goods or services;

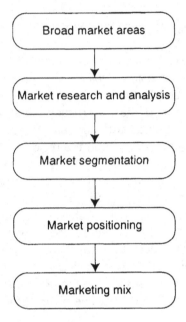

Fig. 7.1 The initial marketing approach

- the establishment of a market position – what policy would be adopted for each market segment;
- the outline of a marketing mix – the marketing mix is the range of marketing activities that can be undertake to address each segment in order to meet the policy objectives.

By November 1994 a draft marketing strategy [44] had been produced for comment using the initial approach.

7.7 A draft marketing strategy

In November 1994, the first draft marketing strategy was produced for comment. At the beginning of the marketing strategy there was a reminder of the underlying assumptions and guiding principles for the project. The first underlying assumption was that all key organizations involved in the centre represented distinct marketing agendas on both the supply- and demand-side of the equation: no single organization would be able to deliver all the aims and expectations of an architecture centre. The implication of the assumption has been that each participating organization has been asked to reappraise its existing role and stress has been placed on determining a common approach and synergy.

The second assumption was that the complex social and economic backdrop to the creation of an end of millennium architecture centre would require a sophisticated marketing response on both the supply and demand side of the equation.

7.7.1 Guiding principles

In order to start from a common point it was felt that there should be a set of groundrules or guiding principles. These guiding principles are listed below:

- The first guiding principle, stated earlier in the chapter, has been the firm conviction that no single professional interest should dominate the marketing agenda or determine the programme of future activity. This principle would determine how the centre addressed certain markets and how the centre would be perceived in both the business sector and the public realm. The marketing strategy had to take account of the need for a non-protective, non-institutional approach.
- The centre should be a focus for open, critical debate. This principle was embodied in the first year programme of activity of the centre and was seen as important if the centre was not to reflect an institutional view.
- The centre should be a 'centre of excellence'. The issues of what service would be on offer and in what areas it would excel were developed in the strategy.
- The activities of the centre should have an international dimension and perspective wherever possible. This is one of the observations that came out of the scenario planning workshop with William Roe and was taken into consideration in the strategy.
- The marketing strategy should be a simple, systematic and sustained programme of continuous communication.
- Procurement was viewed as a pivotal issue to be addressed both qualitatively and quantitatively. The marketing strategy assumed that the ultimate measure of performance of an architecture centre would be its ability to influence the quality and design of the built environment in its own region.
- In the rush to improve the logistics of service and construction, the parallel concerns of design, purpose and value have been ignored. The architecture centre must address these issues.

- The new centre should 'seek to address the real needs of individuals and communities and empower them to play an active role developing their own futures'. This was said by Joan Kean, director of Newcastle Architecture Workshop, in its fifteenth year celebration report. The aim was adopted as one of the Northern Architecture Centre's guiding principles.
- The mission statement of the centre is defined as enabling people to increase their knowledge, understanding and appreciation of contemporary and historical architecture and in so doing raise their levels of awareness and involvement in shaping their future environment.
- All the organizations occupying the centre are regarded as equal partners.

7.7.2 The markets for the Northern Architecture Centre

Four broad markets were identified in the first strategy namely:

- arts, heritage and conservation;
- community and education;
- procurement;
- tourism.

Two approaches had been used in parallel. First, the scenario planning workshop facilitated by William Roe Associates had begun to paint a picture of what the centre would look like in the medium to long term. Second, a synthesis marketing framework had to be drawn up which covered the four markets shown above. The emphasis here was on taking a holistic view of the entity called 'the Northern Architecture Centre' and working out its relationship with the markets listed above and the wider audience of the institutions and organizations making up the local and regional economy.

7.7.3 Arts, heritage and conservation

The Northern Architecture Centre was faced with the task of 'promoting, fostering and encouraging interest in the art of architecture and concern for the aesthetic quality of the built environment. The architecture centre should be in a position to encourage competitions, promote demonstration projects in building and urban design, support designers with innovative ideas and encourage research into design problems' (Gutman [9]).

It was felt that the Centre's involvement and interaction with the arts scene in the region would provide a highly visible and accessible platform for architecture and the built environment. In effect, it would be part of the Architecture Centre's shop window. The Centre should fit into an existing, active arts market in the north-east of England and would be in a position to develop an outreach programme of cultural activity in a cultural catchment area extending to Cumbria and Teesside.

The art of architecture was felt to be the medium through which people would have their first contact and understanding of the built environment and the world of architecture. The architecture as art debate had enjoyed regular national media coverage in recent years and the public had been informed by a programme of investigations into the cultural and aesthetic issues that face the modern world. The result was a strong alignment between art, architecture and the environment which the Northern Architecture Centre would be in a position to exploit.

The comparison and interaction between the possible activities of the architecture centre and the existing activities of Northern Arts (the northern office of the Arts Council) provided an indication of the potential market position of the architecture centre. Several distinct but related observations were made in the draft marketing strategy:

■ Although the understanding and enjoyment of architecture was part of the Northern Arts programme, it nevertheless operated on the periphery of architecture and had no serious remit to cover this area. The Northern Architecture Centre could therefore be better placed to be an advocate of quality architecture.
■ The Northern Architecture Centre could give awards and grants for architectural merit and 'architecture as a fine art'.
■ The architecture centre would be better placed to be a centre of information about art, architecture and the environment.
■ The architecture centre could offer some arts based services. For example, publications coming out of the architecture centre could be endorsed by the participants in the centre.
■ The architecture centre would be in a position to stimulate new areas for arts in the northern region.

7.7.4 Market position: heritage and conservation

The primary aim of existing organizations such as English Heritage was to 'keep buildings in beneficial use' and to 'manage, promote and make interesting' these venues. Much of the management and promotion of heritage activity in general could involve the arts and for this reason arts, heritage and conservation have been included under one heading in the draft marketing strategy. However, the market position of the Architecture Centre differs between 'arts' in one market and 'heritage and conservation' in a second market, despite the close links.

When the draft marketing strategy was written, the steering group was not in a position to say exactly what the relationship of the Architecture Centre should be to the heritage and conservation market. The issues were not as clear cut as the possible interaction between, say, the Northern Arts and the Architecture Centre. One of the complications was the aspect of 'branding'. Or to be more precise, the decision to 'brand' or not to 'brand'. It is worth spending some time explaining what is meant by branding and why it was seen as important.

7.7.5 Branding

Branding is the means by which you can identify goods or services and differentiate them from those of other providers or competitors. This has often been done by using names, terms, signs, symbols or designs. The advantage of branding is that it is a simple, consistent form of communication. For example, the three-pointed star of Mercedes Benz conveys a wealth of perceptions and assumptions to the consumer. The major disadvantage of branding is that it can actually exclude markets and act as a barrier. This is particularly true when you are looking at an organization such as the Architecture Centre, which is appealing to diverse markets with a range of services. In the case of the Architecture Centre, choice and diversity were felt to be as important as branding.

In the heritage and conservation market, English Heritage was highly branded. The positive image of English Heritage was one of opening up sites to the public. The negative perception was of unrealistic altruism that would, for example, allow old buildings to fall down rather than use certain types of window frames. Neither of these statements was totally true or conveyed accurately the wide range of activity of English Heritage. It was felt therefore, that there would be an opportunity for the Architecture Centre to address the heritage and conservation market in a way that would not be possible by an organization such as English Heritage, given the constraints of its statutory obligations and its relatively high level of branding.

The arguments for or against branding have particular relevance to choice of the final name for the centre and it subsequent promotion. There is a school of thought among marketing professionals that names, by themselves, convey very little. For example, in the consumer goods market, the nonsense names of Daz, Omo and Persil cannot by themselves convey anything; but they are all highly branded products. The branding comes from the simplicity and strength of the promotional message. The only exception is a situation in which the name is functional. For example, the title 'Techsperience Centre' has a message about technology and hands on experience in the name. However it would leave little room for manoeuvre if you wanted to address a wider audience. The same argument could be applied to the suggestion that the naming of the centre be subject to a public competition. How would you formulate and present such a diverse and complicated brief to the public? If you narrowed the brief, would you be stuck with a functional title that gave an unexpected and inappropriate branding? Never assume that you know what a word or phrase means to the general public.

Let us look at a simple branding scenario. The decision has been made to adopt the name 'The Northern Architecture Centre'. A striking modern logo has been designed, following a public competition, and the board of directors of the Centre has to make a decision about how the Centre should be promoted in each market. The decision is taken that a single sheet leaflet will be produced for each market and statements will be made about the Centre's position. In effect, the board of directors would have branded the Centre in each sector. In the draft marketing strategy it was recommended therefore that only one market should be branded, namely the market of visitors to the Centre. This was because branding would hinder the diversity, synergy and influence of the Centre. It might even squeeze out constituencies that would have been otherwise attracted to the Centre's position. It was recommended that the marketing in the non-visitor sector should be presented in a portfolio format that avoided outward positioning and concentrated on opportunity and diversity.

7.7.6 Newcastle Architecture Workshop

The Newcastle Architecture Workshop already had a distinct marketing agenda relating to community and education. That agenda was concerned with enablement, empowerment and interpretation. The aims of the workshop were

> to foster and promote the maintenance, improvement and development of knowledge, understanding and appreciation and care of the built environment and its immediate surroundings amongst the inhabitants of the beneficial area. To facilitate the same and establish an architectural

resource centre for the community and provide architecture workshops, educational and exhibition facilities, publications, lectures, meetings instruction and resources as a public service.

The belief of the Newcastle Architecture Workshops brief covered local empowerment linked to local decisions and enablement in the training and education arena. This was a closely defined market position and the the recommendation in the draft marketing strategy was that this position remained unchanged. Newcastle Architecture Workshop would be able to maintain its market differentiation among its client groups and at the same time would be able to benefit from the corporate clout of the architecture centre. Indeed, the newness and freshness of the Architecture Centre was seen as an advantage and might lead to further funding.

7.7.7 The Northern Architectural Association

In the first marketing strategy it was suggested that the Northern Architectural Association reviewed its existing activities in the light of the cultural programme for the Architecture Centre. All of these proposed cultural activities would fall within the Association's wide-ranging Memorandum of Association and in many ways the Northern Architecture Centre would be uniquely positioned to take on the task. For example the Association's distinctive, historical logo (see Fig. 7.2) could be used for branding in the visitor and tourism markets. The marketing implication would be that the Northern Architecture Association would have to be repositioned to take on board a much larger programme of activity.

Fig. 7.2 The Northern Architectural Association logo

7.7.8 The Centre for the Built Environment

The main objectives of the Centre for the Built Environment were to provide:

- an integrated service of education, training and career advice opportunities and course provision across the range of disciplines at all levels;
- a regional focus for the construction industry, its professions and other organizations;
- a centre for cooperation in continuing professional development, consultancy and research;
- a focus for public awareness of the built environment.

In 1994 the Centre for the Built Environment was a relatively new organization and was primarily concerned with marketing taught courses to a wide audience ranging from young people to industrialists. It acted as a marketing agency for scheduled courses targeted at the building industry but in the longer term the centre was planning to extend its activities to cover both education and training across the whole of the region. It was agreed that further work would be needed to look at how, for example, the centre could be involved in stimulating open debate and how it would operate under the umbrella of the Northern Architecture Centre.

7.7.9 Procurement and the regional construction industry

In the preparation of the first marketing strategy, the Northern Architecture Centre received assistance from the Northern Development Company (NDC). The NDC had researched and collated information about the construction sector in the northern region of Cumbria, Northumberland, Tyne and Wear, Durham and Teesside. Here are some of their relevant findings.

A total of 740 manufacturing companies in the region depend on the construction industry for business. These companies can be split into five sectoral classifications, as shown in Table 7.1. The employee numbers are shown in Table 7.2. Table 7.3 gives company turnover.

On the service side there were 715 registered architects working in the region and a greater, unknown number of engineering consultants, quantity surveyors and other consultants operating in the construction industry.

The draft marketing strategy stated that the ultimate measure of

Table 7.1 Company numbers (with duplication)

Installation, fixtures and fittings	353
Construction and repair of buildings	293
Building completion works	253
Civil engineering	152
General construction and demolition	137

Table 7.2

Employment bands (Number of employees)	Number of companies
10–19	260
20–24	57
25–49	186
50–99	108
100–199	55
200–249	11
250–349	23
350–499	20
500–999	13
1000+	7

Table 7.3

Company Turnover	Number of companies
< £50,000	91
£50,001–£100,000	62
£100,001–£500,000	175
£500,001–£1,000,000	119
£1,000,001–£5,000,000	171
£5,000,001–£10,000,000	50
£10,000,001–£20,000,000	20
> £20,000,000	51

performance of an architecture centre would be its ability to influence the quality and design of the built environment in its own region. This could only be achieved through influencing procurement decisions. If the Architecture Centre did not have influence on the procurement process, then its efforts will have been a waste of time. There was good reason for taking such a strong stance. The northern region accounts for just under 5% of GDP. However, it receives nothing like that amount of procurement in any sector. The difference can be expressed in a number of ways. For example, it accounts for tens of thousands of jobs or numerous buildings not built. Both measures are of particular relevance to the construction industry which is the largest single employer in the region (outside the public and service sectors).

The northern region's recorded share of construction output in Great Britain has been consistently below the region's share of gross domestic product (GDP). Historical figures are shown below in Table 7.4.

Although the percentage difference between GB construction output and GB GDP looks small, the actual financial level is considerable. For example, the estimated 0.6% shortfall of construction output in 1994 has a value of approximately £250 million. This amount only reflects the actual level of construction. It does not show the amount of lost turnover of companies that supply other products and services to member organizations of the construction industry. The negative impact on the quality and quantity of buildings in the region is considerable.

The regional construction industry is therefore the sector hardest hit by the procurement shortfall. Construction output is the physical embodiment of the low level of investment and re-investment in the regional sector. The regional construction sector is also small relative to the other

Table 7.4 Comparative share of construction output

Year	Output	Share of GB output	Share of GB GDP
1987	£1,608m	4.7%	5.0%
1988	£1,717m	4.1%	4.9%
1989	£2,059m	4.2%	4.9%
1990	£2,209m	4.3%	4.8%
1991	£2,055m	4.3%	4.9%
1992	£1,942m	4.4%	4.9%

Source : Tyne and Wear Research and Intelligence Unit

sectors in the regional economy, and compared with construction sectors in similar regional economies in the UK. The construction industry is therefore too small in both relative and absolute terms. It was suggested in the draft marketing strategy that the Northern Architecture Centre addresses the regional construction output shortfall through a marketing and cultural consumption programme. The challenge for the Northern Architecture Centre would be to influence, both by itself and with partner agencies, the investment in and quality of the built environment in the region.

7.7.10 The role of the RIBA regional office

The activities of the RIBA at national level had been under review for several years in response to the wide ranging changes in the construction industry and the changing role of design professionals. There was a need therefore to reflect on how the role of the RIBA regional office would change if it was part of the Northern Architecture Centre.

The marketing strategy recognized the need for regional RIBA offices to have greater influence on the procurement process. The Northern Architecture Centre could potentially improve links into the local and regional economic network and provided a marketing platform in a number of ways:

- There could be a matching of projects and resources. The Architecture Centre would have access to an on-line procurement database provided by the Northern Development Company and would have access to site information in the region and funding availability from its synthesis marketing activities. The RIBA regional office could have both access and influence through an updated client advisory service,
- The synthesis marketing efforts of the Northern Architecture Centre would place a great deal of emphasis on project assembly, putting architects in touch with new project development at an early stage. The RIBA regional office could be involved in this process.
- The Northern Architecture Centre would be in a position to develop a family of arguments with other interested groups in a way that would not be possible by an isolated RIBA regional office.
- The Northern Architecture Centre could develop a regional and national lobbying agenda with full RIBA support.
- The regional office of the RIBA would able to contribute to and benefit from the centre's marketing programme.

7.7.11 Tourism and the Northern Architecture Centre

A number of possible locations for the Northern Architecture Centre had been identified on the East Quayside in Newcastle-upon-Tyne for a centre occupying between 1600 and 2000 square metres of space. The East Quayside offered a prime, high profile location with potential to attract visitors and tourists.

The draft marketing strategy estimated that a potential visitor throughput in the first year of operation of 40000 visitors. In order to achieve this target, the Northern Architecture Centre would have to do a number of things, including:

■ integrating the Architecture Centre's tourism marketing with the marketing of Newcastle and its architecture as a regional centre;
■ linking the proposals for conference and training facilities to the visitor attraction elements of the Centre;
■ target what the Northumbria Tourist Board [45] calls the 'sights and service' market segment. This segment has 'the oldest profile and is the most package oriented. It enjoys sightseeing and looks for comfort and service. It is the segment most likely to use hotel accommodation';
■ target relevant travel brochures at least eighteen months ahead of the opening of the Centre;
■ target the travel media about six months ahead of the opening;
■ target the day visitor market segment separately through specialist marketing outlets;
■ develop a strong branding statement incorporating all of the above.

In broad terms, the visitor market segmentation for the Northern Architecture Centre is likely to comprise educational and professional visitors and day trip tourists. The intention of the Northern Architecture Centre would be to target this short break market as a priority and to complement other tourism activities in the city and region.

It was anticipated that educational visitor levels would be high, in the order of one-third of the total. The remaining balance of professional visitors to, for example, conferences was felt to be an underestimate. The total level of visitor throughput does not take into account all the activities of the organizations that would occupy the building and the estimated professional visitor level could be exceeded through careful marketing.

7.8 The synthesis marketing strategy

By the beginning of 1995, the first marketing strategy had begun to resolve at least some of the marketing composition issues; but question marks still hung over the marketing valency and marketing entelechy of the Northern Architecture Centre. What synthesis factors would bind the centre to a number of audiences in the northern region in a long-term enduring relationship? What processes would be required to deliver the services of the new centre to these various audiences? What driving forces would influence the outcome of events in the longer term development of the centre?

The picture that has emerged of the role and function of the Northern Architecture Centre is a body that acts as a marketing platform to:

- promote cultural consumption linked to economic regeneration in a modern economy;
- market the activities of the key partners;
- create a centre for debate;
- market cultural and economic aspects of the region at an international level;
- support the regional construction industry.

It is implicit that in this process that the centre will market the unique contribution that design professionals (architects and engineers) make to the building process. However, the directors of the Northern Architecture Centre have gone to a great deal of trouble to ensure that the marketing strategy takes account of the need for a non-protective, non-institutional approach.

7.9 Culture, community, construction

As more and more people began to get involved in the plans for the Northern Architecture Centre proposal in 1995, there was a need to provide a simple explanation of the operation and positioning of the centre. The headings 'culture, community, construction' were used to encapsulate the activities of the centre. The headings represent:

- the promotion of architecture as the physical manifestation of culture;
- engagement with the business community, the general public, educators and students;
- support for the regional construction industry.

The 'culture, community, construction' headings were the end product of scenario planning and synthesis marketing effort. However, they did not relate directly to the marketing valence, marketing composition and marketing entelechy elements of the synthesis marketing agenda. 'Culture' is only one of a number of synthesis factors that will link the Centre to the various markets in the future. By itself, it is unlikely to result in the level of marketing valence that would ensure long-term success. It is however the starting point for the synthesis marketing activity.

7.10 The cultural programme

In 1996, the northern region hosts the Year of the Visual Arts: an honour won in open competition. It is intended that the Northern Architecture Centre will support, continue and supplement where possible the arts activity that has been highlighted during the year. The Year of the Visual Arts provides a starting point for the long-term development of cultural provision, and Northern Architecture Centre is well placed to support subsequent artistic and cultural activity.

In September 1995 Northern Architecture Centre Ltd commissioned a report [46] by Michael Collier, Cat Newton-Groves and Germaine Stanger. The aim of the report was to look in more detail at the cultural programme. The report concluded that the Northern Architecture Centre should

> run, coordinate (in collaboration with the partners) and actively market a
> programme which:

- informs and challenges people's perception of the urban and rural landscape through a series of exhibitions, critical debates, publications and temporary events.
- is seen to lead the field locally, nationally and internationally in stimulating debate about the role of architecture and the power of people to influence and change their environment.
- examines the culture of the urban and rural landscape in the northern region within the context of national and international developments.

7.11 Engagement with the community

Engagement with the 'community' is seen as the central plank of the synthesis marketing effort. In March 1995, Sir Richard Rogers stated in the

Reith Lectures [47], *Cities for a Small Planet*, that:

> We need to inform and involve citizens in the problems of their environ-
> ment and their city. Citizens' participation needs to be made informal
> and interesting. The Victorians built public libraries, we should build archi-
> tecture centres. The Architecture Centre is where the planning committee
> could meet in public. It would become the focus for public debate on
> strategic plans, planning applications and competitions. It would hold
> lectures, exhibitions, courses and debates about the city and its architec-
> ture. At the very heart of each centre should be adaptable working
> models of the city and the neighbourhood.

As mentioned in Chapter 3, Robert Gutman sees cultural consumption as
a sign of a mature society in which 'appreciation of aesthetic values is no
longer confined to an upper class but extends to many more groups in
society'. The corollary is that cultural consumption should involve public
participation and should take into account public opinion. Cultural
consumption should include public participation in the decision making for
new buildings, refurbishment and the preservation of historic building. The
synthesis marketing strategy for the Northern Architecture Centre must
take into account the need for increased user and community participation as a part of the cultural consumption process. The process of
marketing entelechy for the centre will depend on the quality of
'community' engagement and the matching process in both the public and
educational spheres.

7.12 The second scenario planning workshop (June 1995)

The second half-day scenario planning workshop brought together six
directors of Northern Architecture Centre Ltd, a company limited by
guarantee, which had been incorporated with director level representation
from each of the four key partners. The scenario planning workshop was
again led by William Roe.

The purpose of the workshop was to look again at how the
partners were being melded together, to discuss the wider ownership
of the project by a wider group of stakeholders and find some sense of
the the critical success factors that would be important in the coming
year.

The starting point for the workshop was to identify the predetermined elements, i.e. the fixed points of understanding and 'the factors that
we can count on'. (Note that neither of the scenario planning workshops

stuck rigidly to the process outlined in Chapter 3. The process should be adjusted to suit the composition of the group and the point in time when the scenario planning is taking place.) The predetermined elements included:

- recent grant support of £100 000 from the Arts Council Lottery Board;
- the existence of Northern Architecture Centre Ltd as an up and running company capable of moving the project forward;
- tangible goodwill from many organizations and institutions throughout the region;
- an in principle offer of £450 000 worth of European Regional Development Fund capital monies subject to finding the balance of the funding for a building.

The remainder of the half day was spent covering:

- the benefits of the centre to a number of groups;
- the perceptions of problems that faced the individual partners and the Northern Architecture Centre;
- the likely impact of the centre on the various target markets;
- the short-term (one-year) programme of events and work on the project;
- revenue generation.

The process of scenario planning questions many of the individual assumptions that underpin both organizations and projects. The process can be both helpful and disconcerting. The workshop pulled together a common view and understanding but at the same time provided a reminder of the logistical, funding and human resource problems that faced the Board of Directors of the Northern Architecture Centre. Potential funding from the National Lottery Board offered a large proportion of the capital sum to build a building but offered neither revenue funding for the long-term operation of the Centre nor financial assistance to help secure the required level of running costs for the building. For an organization such as the Northern Architecture Centre, with a diverse set of constituencies, the lack of initial funding to market the centre was a considerable handicap. The scenario planning workshop served as a timely reminder of the critical success factors, including revenue funding, that would need to be in place during the next year.

7.13 Final comment

In the introduction to this book it was stated that if architects and engineers are to face the challenges and changes of the construction marketplace in the next century, they will need to develop a marketing agenda which can be supported and reinforced at every level. Architecture centres are just one level in the overall marketing agenda. They are one of a number of ways in which design professionals can influence their environment and deal with cultural consumption in a modern economy.

The individual success or failure of architecture centres is perhaps less important than the fact that architects and engineers are beginning to find a common purpose and ways of anticipating and responding to events in a rapidly changing world. The forward looking methods and techniques of scenario planning, synthesis marketing and strategic mapping outlined in this book should assist that process.

Bibliography and references

1. Morgan, Neil A. (1991) *Professional Services Marketing*, Butterworth Heinemann.
2. McKenna, Regis (1991) *Relationship Marketing: Own the Market Through Strategic Customer Relationships*, Century Business.
3. Richardson, Brian (1992) 'How architects can promote their services'. *Architects Journal*, 16 December.
4. Grönroos, Christian (1990) *Service Management and Marketing*, Lexington Books.
5. Warne, E. J. D., CB (1992) *Review of the Architects (Registration) Acts 1931–1969*, HMSO.
6. Gide, Charles (1916) *Cours d'économie politique*, authorized translation from the third edition (1913) of the *Cours d'économie politique* under the direction of Professor William Smart by Constance H. M. Archibald MA (1916), Harrap.
7. Hutton, Will (1995) *The State We're In*, Jonathan Cape, London.
8. Walsh, Keiron (1989) *Marketing in Local Government*, Longman, Local Government Training Board.
9. Gutman, Robert (1988) *Architectural Practice: A Critical View*, Princeton Architectural Press.
10. Richardson, Brian (1993) 'The importance of marketing'. *Architects Journal*, 31 March.
11. Greenberg, Stephen (1993) 'Architecture: a professional or commercial enterprise'. *Architects Journal*, August.
12. Bayliss, A. J. (1989) *Marketing for Engineers*, Peter Peregrinus, Institute of Electrical Engineering.

13. Coxe, W. et al. (1992) *Success Strategies for Design Professionals: Superpositioning for Architecture and Engineering Firms*, Krieger, Malabar, FL.
14. Maister, David H. (1993) *Managing the Professional Service Firm*, Free Press, (reprinted from the April 1984 issue of *American Lawyer*).
15. RIBA (1992) *Phase 1: Strategic Overview, Strategic Study of the Profession*, May.
16. Levitt, Theodore (1962) *Innovation in Marketing: New Perspectives for Profit and Growth*, McGraw Hill.
17. Bovée, C. L. and Thill, J. V. (1992) *Marketing*, International Edition, McGraw Hill, New York.
18. Oakland, John S. (1993) *Total Quality Management*, 2nd edn, Butterworth Heinemann.
19. Schwartz, Peter (1992) *The Art of the Long View*, Century Business.
20. Amin, Ash. (ed.) (1994) *Post Fordism: A Reader*, Blackwell.
21. Hutton, Will (1993) 'Three thirds Britain'. First published in *New Times*, May.
22. Hobsbawm, Eric (1994) *Age of Extremes: The Short Twentieth Century, 1914–1991*, Michael Joseph.
23. Hall, Stuart (1988) 'Brave new world'. *Marxism Today* 24–29 October.
24. Hutton, Will (1995) 'High risk strategy: the 30/30/40 society'. *The Guardian*, 30 October.
25. Lewis, Brian J. (1988) 'The globalisation of consulting engineering'. Abstract for the American Society of Engineering Management.
26. Schneider, Eric (1992) 'Segmenting a diverse profession'. *RIBA Strategic Study of the Profession: Phase 1*, May.
27. Smyth, Hedley (1994) *Marketing the City*, E. & F. N. Spon.
28. RIBA (1993) *Strategic Study of the Profession. Phase 2: Clients and Architects*, October.
29. Martin, Christopher, Payne, Adrian and Ballantyne, David (1991) *Relationship Marketing: Bringing Quality, Customer Services and Marketing Together*, Butterworth Heinemann.
30. Peters, Thomas J. and Waterman, Robert H. Jr. (1982) *In Search of Excellence: Lessons from America's Best-run companies*, Harper and Row.
31. Quirk, Randolph (1971) *The Use of English*, fourth impression, Longman.
32. McKean, Charles (1994) *Value or Cost: Scottish Architectural Practice in the 1990s*, RIAS.
33. Luchins, Abraham S. (1942) *Mechanisation in Problem Solving: The*

Effect of Einstellung, Psychological Monographs 54, American Psychological Association, IL.

34. Pickar, Roger L. (1991) *Marketing for Design Firms in the 1990s: Charting a Course in a Changing Marketplace*, American Institute of Architects Press.
35. Ansoff, I. H. (1957) *Strategies for Diversification*, McGraw-Hill.
36. Greiner, L. E (1972) 'Evolution and revolution as organisations grow'. *Harvard Business Review*, July/August, 37–46.
37. Bennett, J. and Jayes, S. (1995) *Trusting the Team: The Best Practice Guide to Partnering in Construction*, Centre for Strategic Studies in Construction, Reading University.
38. Coxe, W. and Hayden, M. (1992) *Towards a New Architectural Practice*, Report for the UIA.
39. Updike, John (1989) *Self Consciousness: Memoirs*, Penguin.
40. MacEwen, Malcolm (1973) 'Taking the profession to the public to provoke them'. *RIBA Journal*, May.
41. Coonan *et al.* (1993) 'Architecture centres'. *Architectural Review*, April.
42. Ellis, Adrian (1994) 'Architecture Centres: A Note on the Feasibility Studies, Commissioned by the Arts Council of England.
43. European Economic Development Services Ltd (1993) *Feasibility Study for a Northern Architecture Centre*, December.
44. Richardson, Brian and Bailey, Tim (1994) *The Northern Architecture Centre: A Draft Marketing Strategy (for Comment)*, November.
45. MEW Research (1995) *Holiday Destination Choice*, for the English Tourist Board.
46. Collier *et al.* (1995) *The Art of Architecture: A Cultural Programme*, commissioned by Northern Architecture Centre.
47. Rogers, Sir Richard (1995) *Cities for a Small Planet*, the 1995 Reith Lectures. Broadcast on BBC Radio 4. 12 March.

Index

Printed in the United States
by Baker & Taylor Publisher Services